Holt
Mathematics
Course 3

Texas Lab
Manual Workbook

ISBN 0-03-092714-5

7 8 9 10   018   09 08

Table of Contents

Holt Mathematics

Holt Mathematics

Holt Mathematics

Holt Mathematics

TEKS 8.2.A

LAB
1-7

Model Solving Equations by Subtracting

Use with Lesson 1-7

Materials needed: algebra tiles, pencil

KEY	REMEMBER:
▢ = 1 ▨ = −1 ▯ = x	It will not change the value of an equation if you subtract the same number from each side.

You can use algebra tiles to help you solve equations.

Activity

To solve the equation $x + 3 = 5$, you need x alone on one side of the equal sign. You can achieve this goal by adding −3 to the left side of the equation. You can perform this operation if you add −3 to the right side as well.

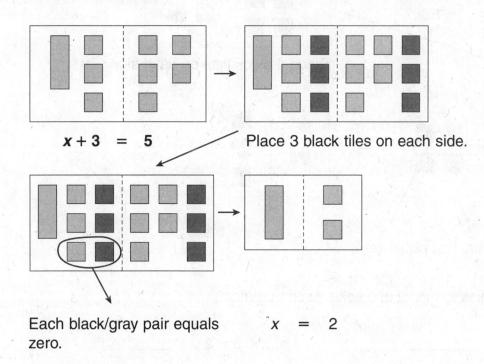

$x + 3 \ = \ 5$ Place 3 black tiles on each side.

Each black/gray pair equals zero. $x \ = \ 2$

Holt Mathematics

TEKS 8.2.A

LAB 1-7

Model Solving Equations by Subtracting

Use with Lesson 1-7

1. Use algebra tiles to model and solve each equation.

a. $x + 2 = 4$ _____

b. $x + 3 = 5$ _____

c. $x + 1 = 5$ _____

d. $5 + x = 6$ _____

The same technique works to model the solution to the equation $x + 4 = 1$, where there are fewer tiles on the right side than the left.

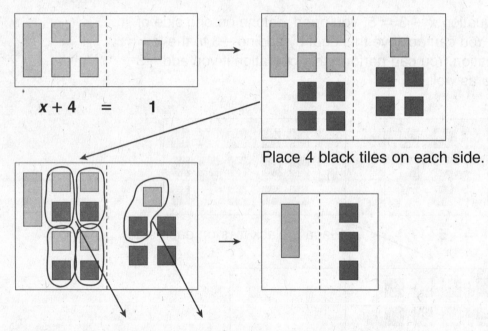

$x + 4 = 1$

Place 4 black tiles on each side.

Remove all pairs that equal zero.　　　$x = -3$

2. Use algebra tiles to model and solve each equation.

a. $x + 4 = 3$ _____

b. $x + 5 = 2$ _____

c. $x + 3 = 1$ _____

d. $x + 3 = 3$ _____

Holt Mathematics

Name _____ Date _____ Class _____

TEKS 8.2.A

Model Solving Equations by Subtracting

Use with Lesson 1-7

Think and Discuss

1. How do you know how many black tiles to place on each side?

   _____

   _____

2. Why is it important to add the same number of black tiles to both sides of the model?

   _____

   _____

3. Why are you allowed to remove pairs of black and gray tiles from the model?

   _____

4. How is solving the equation $x + 1 = 3$ different from solving $x + 3 = 1$? How is it the same?

   _____

   _____

   _____

   _____

Try This

Use algebra tiles to model and solve each equation.

1. $x + 1 = 2$ _____   2. $x + 3 = 5$ _____   3. $x + 2 = 7$ _____

4. $x + 3 = 2$ _____   5. $x + 4 = 2$ _____   6. $x + 2 = 1$ _____

7. $x + 1 = -1$ _____   8. $x + 1 = 0$ _____   9. $x + 2 = -1$ _____

Holt Mathematics

Name _____ Date _____ Class _____

Model Solving Equations by Dividing

LAB
1-8

Use with Lesson 1-8

Materials needed: algebra tiles, pencil

KEY	REMEMBER:
▢ = 1 ■ = −1 ▯ = x	It will not change the value of an equation if you divide both sides by the same number.

You can use algebra tiles to help you solve equations.

Activity

To solve the equation $2x = 4$, you need x alone on one side of the equal sign. You can achieve this goal by dividing the left side of the equation by 2. You can divide as long as you divide the same way on both sides.

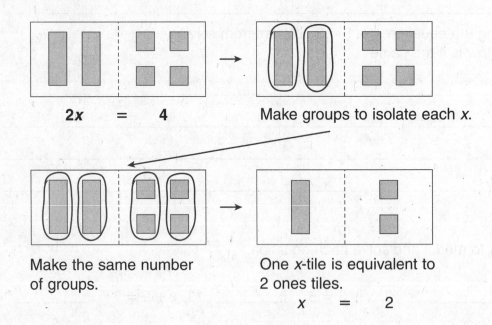

$2x = 4$ Make groups to isolate each x.

Make the same number of groups.

One x-tile is equivalent to 2 ones tiles.

$x = 2$

Holt Mathematics

TEKS 8.2.A

  Model Solving Equations by Dividing

Use with Lesson 1-8

1. Use algebra tiles to model and solve each equation.

a. $2x = 8$ _____

b. $3x = 9$ _____

c. $4x = 8$ _____

d. $2x = 6$ _____

Solving the equation $2x = -2$ is not much different. You can use tiles to represent negative numbers just as you can use them for positive numbers. Once you set up the equation using the x-tiles and the negative tiles, the method for solving is the same.

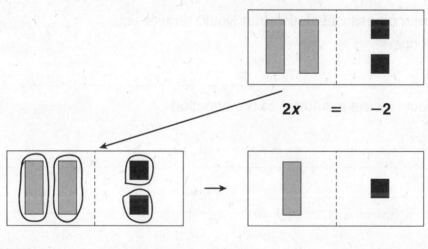

$2x = -2$

Make equal groups on each side. $x = -1$

2. Use algebra tiles to model and solve each equation.

a. $3x = -6$ _____

b. $4x = -4$ _____

c. $2x = -8$ _____

d. $3x = -9$ _____

Holt Mathematics

TEKS 8.2.A

Model Solving Equations by Dividing

LAB 1-8

Use with Lesson 1-8

Think and Discuss

1. When you group tiles, how do you know how many groups to make?

_____

_____

2. How is solving the equation $2x = 4$ different from solving $2x = -4$? How is it the same?

_____

_____

3. Write an equation with a positive solution that would require you to make 3 equal groups.

_____

4. Explain how you would model a solution to the equation $2x = -6$.

_____

_____

_____

Try This

Use algebra tiles to model and solve each equation.

1. $2x = 6$ _____

2. $3x = 6$ _____

3. $4x = 12$ _____

4. $2x = -8$ _____

5. $4x = -4$ _____

6. $4x = -12$ _____

7. $3x = -6$ _____

8. $2x = 8$ _____

9. $4x = -8$ _____

Holt Mathematics

Name _____ Date _____ Class _____

TEKS 8.4

LAB 2-3

Adding and Subtracting Rational Numbers

Use with Lesson 2-3

Activity

Susi is a dolphin who likes to jump. She jumps $5\frac{1}{2}$ feet out of the water. Then she falls back into the water, covering a total distance of $9\frac{1}{4}$ feet. How far is Susi now below the water's surface?

Draw Susi's jumps.

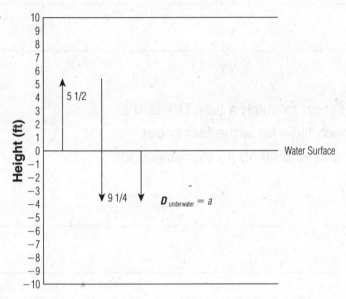

The graph shows that the dolphin is $3\frac{3}{4}$ feet under water.

Check your graph by writing and evaluating an expression for Susi's jumps:

$$D_{underwater} = 5\frac{1}{2} + \left(-9\frac{1}{4}\right) = \frac{11}{2} - \frac{37}{4} = \frac{22}{4} - \frac{37}{4} = -\frac{15}{4} = -3\frac{3}{4}$$

Holt Mathematics

TEKS 8.4

Adding and Subtracting Rational Numbers

LAB 2-3

Use with Lesson 2-3

Think and Discuss

Susi continues to dive 50 more feet to the bottom of the ocean. How deep is the ocean and how do you know?

_____

_____

_____

Try This

A kingfisher dives from a height of 30 feet to catch a fish. The bird's dive covers a total distance of $40\frac{1}{3}$ feet. How far is the fish under water? Check your graph by writing and evaluating an expression for the kingfisher's dive.

_____

_____

_____

Holt Mathematics

Name _____ Date _____ Class _____

Equations, Tables, and Graphs

LAB 3-5

Use with Lesson 3-5

Relationships can be represented as equations, tables, or graphs. Each representation shows the same data, but in a different way.

Materials needed: tokens, ruler
Activity

At the Sunshine Arcade, you need 2 tokens for a minute of game time. Use your tokens and the boxes below to model the relationship between the number of tokens you have and the number of minutes you can play your favorite game.

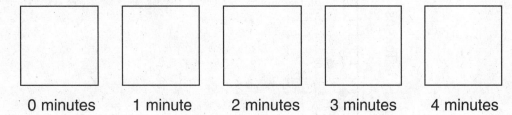

0 minutes   1 minute   2 minutes   3 minutes   4 minutes

How many tokens do you need to play the game for 0 minutes? None! Leave the box above 0 minutes empty.

How many tokens do you need to play the game for 1 minute? Since you need 2 tokens for a minute of game time, stack 2 tokens in the box above 1 minute.

You need 4 tokens to play for 2 minutes. Stack 4 tokens in the box above 2 minutes.

Now, stack the number of tokens you need to play for 3 minutes and for 4 minutes.

You can show the relationship between the number of minutes and the number of tokens in a table. Complete the table as shown. Use the table to predict the number of tokens you need to play for 5 minutes.

Notice that the number of tokens is twice the number of minutes. You can write an equation to relate the number of tokens (t) to the number of minutes (m).

Minutes (m)	Tokens (t)
0	0
1	2
2	
3	
4	

$$t = 2m$$

Holt Mathematics

Name _____ Date _____ Class _____

Equations, Tables, and Graphs

Use the equation to find the number of tokens you need to play for 5 minutes. Compare your result with the prediction you make using the table. They should be the same!

The number of tokens needed for 0, 1, and 2 minute(s) are shown on the graph below. Plot points for the number of tokens needed for 3 and 4 minutes.

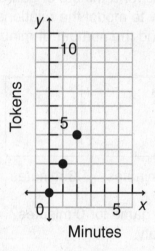

Use a ruler to draw a line connecting the points. Extend your line up to the right. Based on the graph, how many tokens do you need for 5 minutes? Your answer should be 10, the same answer you got using the table and the equation!

Think & Discuss

1. How do you know that the table, equation, and graph all represent the same relationship?

_____

_____

2. How would you find the number of tokens you need to play for 25 minutes? Explain.

_____

_____

Holt Mathematics

Name _____ Date _____ Class _____

TEKS 8.5.A

Equations, Tables, and Graphs

LAB 3-5

Try This

A factory can make 25 baseballs in a minute. Make a table, graph, and write an equation to show this relationship. How many baseballs can the factory make in 5 minutes?

Minutes (*m*)	Baseballs (*b*)

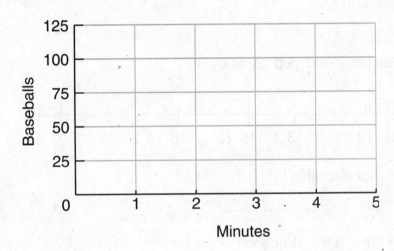

_____

_____

_____

Equation: _____

Baseballs: _____

Holt Mathematics

Name _____ Date _____ Class _____

LAB 3-6

Arithmetic Sequences

Use with Lesson 3-6

A sequence is an ordered list of numbers or objects. In an arithmetic sequence, the difference between one term and the next is always the same. This difference is called the common difference, and it is added to each term to get the next term.

Activity

Look at the number line below. Start at 2 and skip count by 3s.

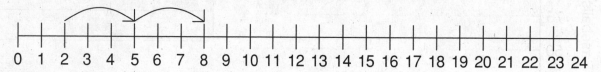

0 1 2 3 4 5 6 7 8 9 10 11 12 13 14 15 16 17 18 19 20 21 22 23 24

Write the sequence you made by skip counting.

2,   5,   8, _____, _____, _____, _____, _____, . . .

Find the difference between each term in the sequence.

2, 5, 8, 11, 14, 17, 20, 23, . . .

∨∨∨∨∨∨∨

+3, +3, __ __ __ __ __

The sequence is arithmetic, and the common difference is 3.

You can write a formula to find the terms in an arithmetic sequence. If y represents a number in the sequence and n represents the number's position in the sequence, then

$y = (\text{COMMON DIFFERENCE})n + (\text{FIRST TERM} - \text{COMMON DIFFERENCE})$.

Find the formula that represents the sequence above:

$y = (\text{COMMON DIFFERENCE})n + (\text{FIRST TERM} - \text{COMMON DIFFERENCE})$

$y = $ _____ n $+$ (_____ $-$ _____)

$y = $ _____ n $+$ (_____)

The formula is $y = 3n + (-1)$, which you can write as $y = 3n - 1$.

What was the 5th term in the sequence? Use the formula to find y when $n = 5$. The answer is the same.

Holt Mathematics

TEKS 8.16.A

 Arithmetic Sequences

LAB
3-6

Think & Discuss

1. How can you tell if a sequence is arithmetic?

_____

_____

2. Give an example of a sequence that does not have a common difference. Is this sequence arithmetic? Explain.

_____

_____

_____

Try This

Decide whether each sequence is arithmetic. If the sequence is arithmetic, find the common difference.

1. 5, 10, 15, 20, . . . _____

2. 1, 2, 4, 8, . . . _____

3. −4, 2, 8, 14, . . . _____

4. 6, 2, −2, −6, . . . _____

Use the number line below to answer the questions that follow.

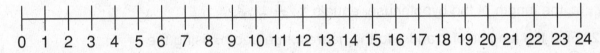

5. Begin at 5 and skip count by 2s. What are the first 10 terms in the sequence? _____

6. What is the common difference? _____

7. Write a formula to find the terms in the sequence.

_____

8. What is the 50th term in the sequence? _____

Holt Mathematics

Name _____ Date _____ Class _____

LAB
4-7

The Real Numbers

Use with Lesson 4-7

Materials needed: paper, pencil
Activity
Mark the length of each side of the triangles by placing a piece of
paper along the side and making a mark at each end of the segment.
Be sure to label each pair of marks with the side they represent. Then,
for each side, place one of the marks on 0 on the number line, and
place a point on the number line that is even with the other mark.

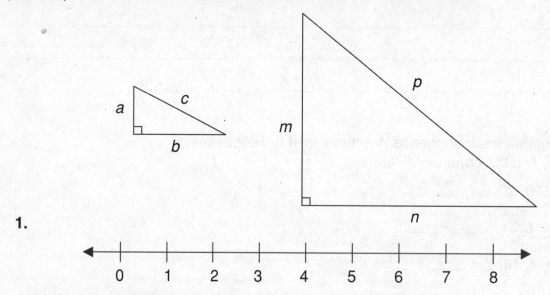

1.

Notice that the lengths of *a, b, m,* and *n* are whole numbers, but
the lengths *c* and *p* are not. The Pythagorean Theorem states
that if *a* and *b* are the lengths of the legs of a right triangle, then
c, the length of the hypotenuse, equals $\sqrt{a^2 + b^2}$.

2. Find the lengths of *c* and *p* to the nearest hundredth, using the
Pythagorean Theorem:

_____

_____

_____

_____

These decimal numbers do not terminate or repeat like rational
numbers do. Therefore, they are known as *irrational numbers*.
Rational numbers and irrational numbers together are known as
real numbers.

Holt Mathematics

Name _____ Date _____ Class _____

TEKS 8.1.C

The Real Numbers

LAB 4-7

Think and Discuss

1. Can a number that can be written as a fraction ever be an irrational number? Explain.

   _____

   _____

2. How could you use a right triangle to graph $\sqrt{2}$ on a number line?

   _____

   _____

   _____

Try This

Mark the lengths of each side of the triangle on the number line. Then use the Pythagorean Theorem to find the length of the hypotenuse of the triangle. Is the hypotenuse a rational or irrational number? Why?

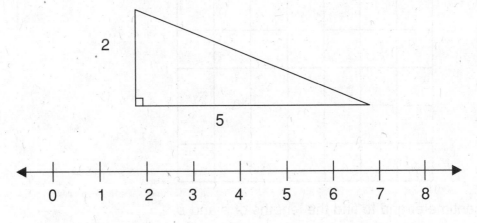

_____

_____

_____

Holt Mathematics

Name _____ Date _____ Class _____

LAB 4-8

The Pythagorean Theorem

Use with Lesson 4-8

Materials needed: centimeter ruler

The Pythagorean Theorem says that in a right triangle, if the lengths of the legs are *a* and *b* and the length of the hypotenuse is *c*, then $a^2 + b^2 = c^2$.

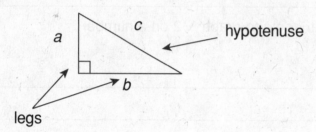

Activity

Use the centimeter grid below to draw a right triangle. Make sure the lengths of the legs are whole numbers. Label the legs *a* and *b*, and label the hypotenuse *c*.

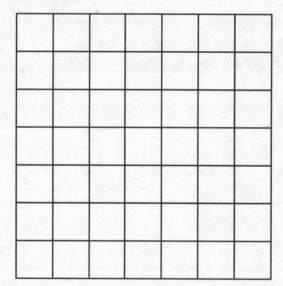

Use the centimeter grid to find the lengths of *a* and *b*.

   a = _____   *b* = _____

Use a ruler to find the length of the hypotenuse to the nearest tenth of a centimeter.

   c = _____

Holt Mathematics

TEKS 8.7.C

The Pythagorean Theorem

LAB 4-8

Use your measurements to find $a^2 + b^2$ and c^2.

$$a^2 + b^2 = \underline{\hspace{1cm}}^2 + \underline{\hspace{1cm}}^2$$

$$= \underline{\hspace{1cm}} + \underline{\hspace{1cm}}$$

$$= \underline{\hspace{1cm}}$$

$$c^2 = \underline{\hspace{1cm}}^2$$

$$= \underline{\hspace{1cm}}$$

Notice that the value of $a^2 + b^2$ and the value of c^2 are almost the same. If it were possible to measure the length of the hypotenuse exactly, these values would be exactly the same.

Think & Discuss

1. If you have a right triangle and you know the length of the hypotenuse and one leg, how can you find the length of the other leg?

_____

_____

_____

2. Is the hypotenuse always the longest side of a right triangle? Explain.

_____

_____

Try This

Use the centimeter grid to draw a right triangle where both legs are 2 centimeters long.

1. Using a centimeter ruler, what is the length of the hypotenuse to the nearest tenth of a centimeter? _____

2. Using the Pythagorean Theorem, what is the length of the hypotenuse, to the nearest thousandth? _____

Holt Mathematics

Name _____ Date _____ Class _____

TEKS 8.16.B

Ratios and Proportions

Use with Lesson 5-1

A ratio is a comparison of two quantities by division. If two ratios make the same comparison, they are equivalent and form a proportion.

Activity
If two ratios are equivalent, then their cross products are the same.

For example, $\frac{1}{2}$ and $\frac{5}{10}$ are equivalent ratios, so $\frac{1}{2} = \frac{5}{10}$.

Let's check the cross products:

$$1 \times 10 \quad \frac{1}{2} \diagtimes \frac{5}{10} \quad 2 \times 5$$

Notice that $1 \times 10 = 10$ and $2 \times 5 = 10$. Since the ratios are equivalent, their cross products are the same.

Suppose $\frac{a}{b} = \frac{c}{d}$. Show that $a \times d = b \times c$.

1. $\dfrac{a \times \square}{b} = \dfrac{c \times \square}{d}$

> Use the Multiplication Property of Equality to multiply both sides of the equation by b.

2. $\dfrac{a \times b \times \square}{b} = \dfrac{c \times b \times \square}{d}$

> Use the Multiplication Property of Equality to multiply both sides of the equation by d.

3. $\dfrac{a \times b \times d}{b} = \dfrac{c \times b \times d}{d}$

> Simplify each fraction. Cross out the common factors.

4. $\square \times \square = \square \times \square$

> You have shown that the cross products are equal!

Holt Mathematics

TEKS 8.16.B

Ratios and Proportions

Think & Discuss

How did you show that equivalent ratios have cross products that are equal?

_____

_____

_____

Try This

Use cross products to determine whether the ratios are equivalent.

1. $\frac{3}{8}$ and $\frac{9}{24}$ _____

2. $\frac{2}{3}$ and $\frac{4}{9}$ _____

3. $\frac{12}{15}$ and $\frac{8}{10}$ _____

4. $\frac{15}{21}$ and $\frac{10}{14}$ _____

5. $\frac{6}{7}$ and $\frac{14}{12}$ _____

6. $\frac{4}{5}$ and $\frac{5}{6}$ _____

Holt Mathematics

Name _____ Date _____ Class _____

TEKS 8.16.A

Similar Figures

Use with Lesson 5-5

In similar polygons, corresponding angles are congruent and the lengths of corresponding sides form equivalent ratios.

Materials needed: protractor, inch ruler, white paper, colored paper, scissors, tape

Activity

If you use small triangles that are the same size and shape to make a large triangle, will the large triangle be similar to the small triangle?

Trace the triangle below 4 times on white paper.

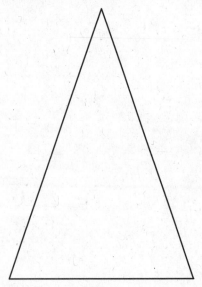

Use your scissors to cut out the 4 triangles you drew. Arrange your 4 triangles without overlaps or gaps to make a large triangle. Tape it to the colored paper.

Your large triangle might look like this:

Now, let's check to see whether each small triangle you traced is similar to the large triangle you made.

Holt Mathematics

TEKS 8.16.A

⬟ LAB **5-5A**

Similar Figures

1. Use the protractor to measure the 3 angles of the small triangle. Record the measure of each angle. Now use the protractor to measure the 3 angles of the large triangle. Record the measure of each angle. What do you notice about the measures of the corresponding angles?

_____

2. Now, measure the lengths of the 3 sides of one of the small triangles to the nearest inch. Record the measure of each side. Measure the lengths of the 3 sides of the large triangle. Record the measure of each side. Write 3 ratios that relate the side lengths of the small triangle to the corresponding side lengths of the large triangle.

☐	☐	☐
☐	☐	☐

3. Are the ratios equivalent?

_____

4. Are the small and large triangles similar?

_____

Think and Discuss

1. Suppose you were to use the large triangle you made in the activity to make a larger triangle in the same manner. Would the new triangle be similar to the large triangle from the activity? Would the new triangle be similar to the small triangle from the activity? Explain.

_____

_____

_____

_____

Holt Mathematics

TEKS 8.16.A

Similar Figures

LAB 5-5A

Try This

1. Draw a rectangle. Can you use your rectangle to make a large rectangle that is similar? Explain.

_____

_____

_____

Name _____ Date _____ Class _____

TEKS 8.16.B

Similar Figures

Use with Lesson 5-5

In scale drawings of composite figures, any polygons in one drawing are similar to polygons in the other drawing. To show that two polygons are similar, you must show that corresponding angles are congruent and the lengths of corresponding sides form equivalent ratios.

Materials needed: protractor, centimeter ruler
Activity
An architect drew two scale drawings of the same building. Show that the scale drawings are similar.

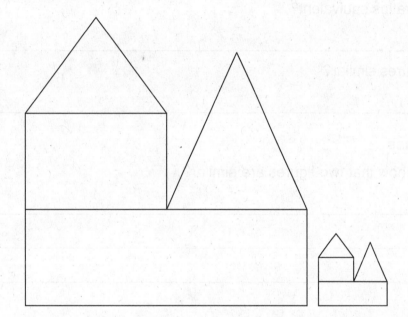

1. How many angles do you need to measure in each drawing?

_____

2. Use a protractor to measure the angles. Label the measures on the drawings.

   Are the corresponding angles congruent? _____

3. How many line segments do you need to measure in each drawing?

_____

Holt Mathematics

TEKS 8.16.B

Similar Figures

LAB
5-5B

4. Use a ruler to measure each line segment to the nearest centimeter. Label the measures on the drawings.

Write 8 ratios that relate the sides of the large drawing to the corresponding sides of the small drawing.

| | | | | | | | |

Are these ratios equivalent?

_____

5. Are the figures similar?

_____

Think & Discuss

How can you show that two figures are similar?

_____

_____

_____

Holt Mathematics

TEKS 8.16.B

Similar Figures

LAB 5-5B

Try This

A photographer gave a rectangular picture to a newspaper. The photographer's picture is 6 inches wide and 8 inches long. When he saw the picture in the newspaper, it was 5 inches wide and 7 inches long.

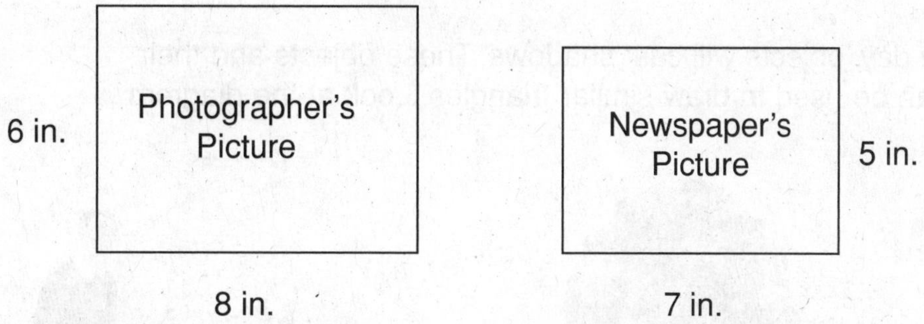

6 in.

Photographer's Picture

8 in.

Newspaper's Picture

5 in.

7 in.

Are the pictures similar? Explain.

_____

_____

_____

_____

Holt Mathematics

Name _____ Date _____ Class _____

TEKS 8.14.D

Similar Figures

LAB
5-5C

Use with Lesson 5-5

In similar polygons, corresponding angles are congruent and the lengths of corresponding sides form equivalent ratios.

Materials needed: tape measure, sunny day, tall tree or flagpole
Activity
On a sunny day, objects will cast shadows. These objects and their shadows can be used to draw similar triangles. Look at the diagram below.

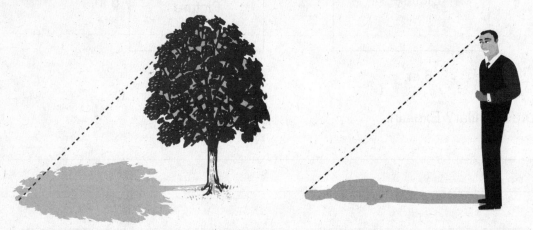

Work with a partner. Use a tape measure to find your height to the nearest inch.

Your height: _____

Now, go outside. Measure the length of your shadow to the nearest inch.

Your shadow: _____

Choose a tall object on your school grounds. Measure the length of its shadow to the nearest inch.

Object's shadow: _____

Holt Mathematics

TEKS 8.14.D

⭐ **LAB 5-5C**

Similar Figures

Write a ratio that relates the length of your
shadow to the length of the object's shadow.

☐
────
☐

Write a ratio that relates the length of your
height to the object's height. Use x to represent
the object's height.

☐
────
☐

Since the triangles formed by objects and their shadows are similar,
the ratios of their corresponding lengths are equivalent. So,

$$\frac{\boxed{}}{\boxed{}} = \frac{\boxed{}}{\boxed{}}$$

Now, find the cross products and solve for x, the object's height.

About how tall is the object?

_____

Compare your answer with other students in your class who chose
the same object. What do you notice?

_____

Holt Mathematics

TEKS 8.14.D

LAB
5-5C

Similar Figures

Think & Discuss

How can you find the height of a building without measuring it?

_____

_____

_____

_____

Try This

1. On a sunny day, a flagpole is casting a 9-foot shadow. Lisa is casting a 3-foot shadow. If Lisa is 4 feet tall, about how tall is the flagpole?

_____

2. A tree is 68 feet tall. Mark is 5 feet tall, and he is casting a 7-foot shadow. How long is the tree's shadow?

_____

Holt Mathematics

Name _____ Date _____ Class _____

TEKS 8.6.A, 8.6.B

 LAB 5-6 # Dilations

Use with Lesson 5-6

A dilation is a transformation that changes the size, but not the shape of a figure. The scale factor describes how much the figure is enlarged or reduced.

Materials: ruler
Activity
Triangle *ABC* is shown on the coordinate plane. Use the following steps to dilate the triangle by a factor of 3, with the origin as the center of dilation.

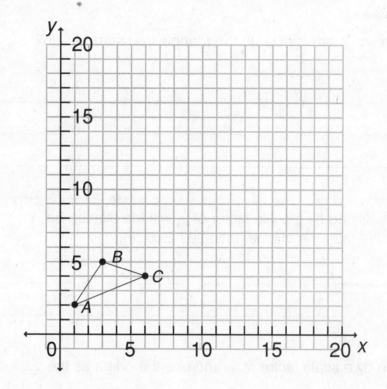

1. The coordinates of the vertices of the triangle are *A*(_____, _____), *B*(_____, _____), and *C*(_____, _____).

Holt Mathematics

LAB 5-6 **Dilations**

2. To find the coordinates of the vertices of the dilation, multiply each coordinate by the scale factor.

$A(1, 2) = (1 \times$ _____ $, 2 \times$ _____ $) = (3,$ _____ $)$

$B(3, 5) = ($ _____ $\times$ _____ $,$ _____ $\times$ _____ $) = ($ _____ $,$ _____ $)$

$C(6, 4) = ($ _____ $\times$ _____ $,$ _____ $\times$ _____ $) = ($ _____ $,$ _____ $)$

3. Now, plot these points on the graph. Use a ruler to connect your points.

Think and Discuss

1. Is the dilation you made similar to the triangle you started with? Explain.

_____

_____

_____

_____

2. If a triangle is dilated by a scale factor of $\frac{1}{3}$, will the dilation be larger or smaller than the original triangle? Explain.

_____

_____

Try This

Dilate the triangle by a scale factor of $\frac{1}{2}$, and use the origin as the center of dilation.

_____

_____

Holt Mathematics

Name _____ Date _____ Class _____

TEKS 8.16.B

Validating Conclusions About Congruence

Use with Lesson 7-6

Materials needed: pencil, inch ruler, protractor

Activity

In order to conclude that two triangles are congruent, you must show that three pairs of corresponding sides have the same length and that three pairs of corresponding angles have the same measure.

1. Use a ruler and protractor to measure the sides of the triangles to the nearest half inch and the angles to the nearst degree.

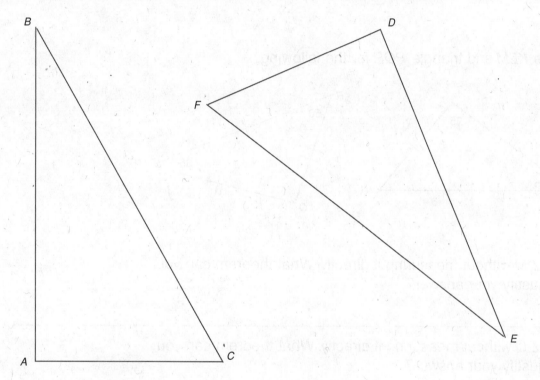

Length of $\overline{AB}$ = _____
Length of $\overline{BC}$ = _____
Length of $\overline{CA}$ = _____
m∠A = _____
m∠B = _____
m∠C = _____

Length of $\overline{DE}$ = _____
Length of $\overline{EF}$ = _____
Length of $\overline{FD}$ = _____
m∠D = _____
m∠E = _____
m∠F = _____

Holt Mathematics

TEKS 8.16.B

LAB 7-6

Validating Conclusions About Congruence

2. Which pairs of sides are corresponding sides?

_____

3. Which pairs of angles are corresponding angles?

_____

4. What can you conclude about the two triangles? Why?

_____

_____

Try This

Use triangle *KLM* and triangle *PQR* for the following.

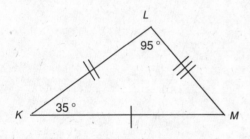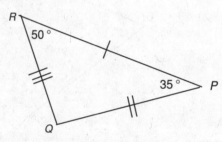

1. Find m∠*M* without measuring it directly. What theorem can you
   use to justify your answer?

_____

2. Find m∠*Q* without measuring it directly. What theorem can you
   use to justify your answer?

_____

3. What can you conclude about triangle *KLM* and triangle *PQR*?
   Why?

_____

_____

Holt Mathematics

Name _____ Date _____ Class _____

TEKS 8.10.A

Perimeter and Area of Rectangles and Parallelograms

Use with Lesson 8-1

Materials needed: $\frac{1}{4}$-inch graph paper, pencil
Activity

1. Use graph paper that is 4 squares per inch. Near the upper left corner of your paper, draw a rectangle that is 4 squares long and 2 squares high. Label this rectangle *A*. Find the perimeter and area of this rectangle.

_____

_____

2. Extend the left side of rectangle *A* down 2 squares. From that point, draw a straight line to the right for 8 squares, then up for 4 squares, then to the left until it meets the upper right corner of rectangle *A*. Label the larger rectangle, which also contains rectangle *A*, *B*. Find the perimeter and area of this rectangle.

_____

_____

Holt Mathematics

TEKS 8.10.A

Perimeter and Area of Rectangles and Parallelograms

LAB
8-1

Note that rectangle *B* is **twice** as long as rectangle *A*.
Note that rectangle *B* is **twice** as tall as rectangle *A*.

3. How much larger is the perimeter of rectangle *B* than rectangle *A*?

_____

4. How much larger is the area of rectangle *B* than rectangle *A*?

_____

_____

Think and Discuss

1. Why is the area of rectangle *B* four times larger than rectangle *A* when the sides are only doubled?

_____

_____

2. If the sides of rectangle *A* are tripled, what do you expect will happen to the perimeter and area of the new rectangle? Justify your answers.

_____

_____

_____

_____

Holt Mathematics

Name _____ Date _____ Class _____

LAB 8-1 Perimeter and Area of Rectangles and Parallelograms

Try This

Below are two parallelograms, *A* and *B*. The sides of *B* are twice the length of the sides of *A*.

1. Calculate the perimeters of *A* and *B* and compare them.

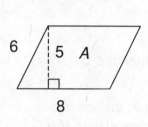

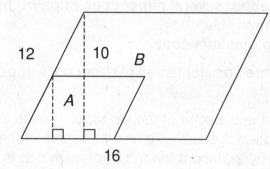

_____

_____

_____

2. Notice that the height of *B* is also twice the height of *A*. Calculate the areas of *A* and *B* and compare them.

_____

_____

3. What can you conclude about the perimeters and areas of rectangles and parallelograms when the lengths of their sides are changed proportionally?

_____

_____

Holt Mathematics

Name _____ Date _____ Class _____

LAB
8-6
Volume of Pyramids and Cones

Use with Lesson 8-6

Volume is a measure of the number of cubic units needed to fill a given space.

Materials needed: several paper cone cups of different sizes, rice, pencil, inch ruler
Activity 1: Volume of a cone.

1. Choose three different sizes of paper cone cups provided by your teacher.

2. Measure the diameter of the top of each cone cup in inches. Be as accurate as possible. To find the radius of the cone divide the diameter by 2. Record the radius of each cup to the nearest eighth of an inch in the table.

3. Next, measure the height of each cup. Record the height of each cup in the table.

4. Fill each cup with rice. Then carefully pour the contents of each cup into a measuring cup. Record the volume of the rice from each cup in cups in the table below and use the formula 1 cup = 14.4 in^3 to find the volume of each cup.

Cone	Radius	Height	Amount of Rice	Calculated Volume (1 cup = 14.4 in^3)
1				
2				
3				

Materials needed: several hollow pyramids, rice, pencil, inch ruler
Activity 2: Volume of a pyramid.

1. Choose three different sizes of pyramids provided by your teacher.

2. Measure the length and width of the base of each pyramid in inches. Be as accurate as possible. Record the length and width of each base to the nearest eighth of an inch. Then calculate the area of the base, *B*, by multiplying the length and the width. Enter your data in the table.

3. Next, measure the height of each pyramid. Record the height of each pyramid in the table.

Holt Mathematics

TEKS 8.8.B

Volume of Pyramids and Cones

LAB 8-6

4. Fill each pyramid with rice. Then carefully pour the contents of each pyramid into a measuring cup. Record the volume of the rice from each pyramid in cups and use the formula 1 cup = 14.4 in^3 to find the volume of each pyramid.

Pyramid	Length of Base	Width of Base	Area of the Base, *B*	Height	Amount of Rice	Calculated Volume (1 cup = 14.4 in^3)
1						
2						
3						

Think and Discuss

1. Calculate the volume of each cone in cubic inches. Use the formula $V = \frac{1}{3}\pi r^2 h$.

2. Compare your calculations to the volume obtained by using rice and the measuring cup.

_____

_____

_____

3. Calculate the volume of each pyramid in cubic inches. Use the formula $V = \frac{1}{3}Bh$.

4. Compare your calculations to the volume obtained by using rice and the measuring cup.

_____

_____

_____

Holt Mathematics

TEKS 8.8.B

LAB
8-6

Volume of Pyramids and Cones

Try This

Find the volume of each cone or pyramid. Use 3.14 for π.

1. _____

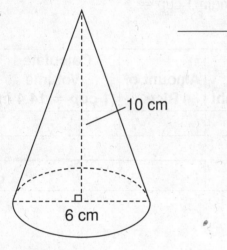

10 cm

6 cm

2. _____

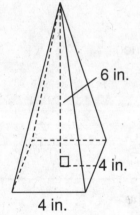

6 in.

4 in.

4 in.

L38

Holt Mathematics

Name _____ Date _____ Class _____

Lateral and Total Surface Areas of Food Packages

LAB 8-8A

Use with Lesson 8-8

Materials needed: pencil, scissors, ruler, package in the shape of a rectangular prism (e.g., box of crackers), package in the shape of a pyramid (e.g., candy box), package in the shape of a cylinder (e.g., soup can)
Activity
In this lab you will measure food packages to find their lateral and total surface areas.

1. **a.** Open and flatten the package that is shaped like a rectangular prism.

_____

b. Measure the dimensions of the lateral faces and the bases. Use the measurements to calculate the area of each face and fill in the table.

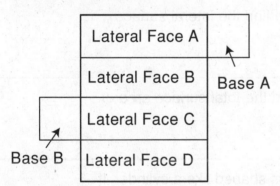

Face	Area
Lateral Face A	
Lateral Face B	
Lateral Face C	
Lateral Face D	
Base A	
Base B	

c. Add the areas of the lateral faces to find the lateral surface area of the prism.

_____

d. Add the areas of all the faces to find the total surface area of the prism.

_____

Holt Mathematics

TEKS 8.8.A

 LAB 8-8A # Lateral and Total Surface Areas of Food Packages

2. a. Open and flatten the package that is shaped like a pyramid.

_____

b. Measure the dimensions of the lateral faces and the base. Calculate the area of each face and fill in the table.

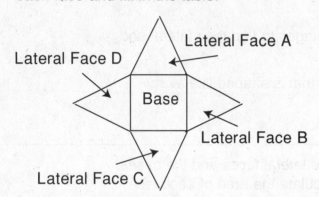

Face	Area
Lateral Face A	
Lateral Face B	
Lateral Face C	
Lateral Face D	
Base	

c. Add the areas of the lateral faces to find the lateral surface area of the pyramid.

_____

d. Add the areas of all the faces to find the total surface area of the pyramid.

_____

3. a. Open and flatten the package that is shaped like a cylinder. If it is a can, carefully peel off and flatten the label.

_____

b. Measure the dimensions of the lateral surface and the bases. Calculate the area of each surface and fill in the table.

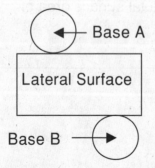

Surface	Area
Base A	
Base B	
Lateral Surface	

Holt Mathematics

★ LAB 8-8A **Lateral and Total Surface Areas of Food Packages**

c. What is the lateral surface area of the cylinder?

_____

d. Add the areas of all the surfaces to find the total surface area of the cylinder.

_____

Try This

1. **Critical Thinking**   Explain how you could find the lateral surface area of a cylindrical soup can without first removing and flattening the label.

_____

_____

_____

TEKS 8.8.A

Using Models to Find Lateral and Total Surface Areas

LAB 8-8B

Use with Lesson 8-8

Materials needed: pencil, ruler, a set of models of geometric solids

Activity

In this lab you will measure three-dimensional figures. Then you will use the measurements to help you find the lateral and total surface areas of each figure.

1. a. Find a model of a rectangular prism in the set of geometric solids.

    b. Measure the length, width, and height of the model. Record the measurements in the figure.

    c. Calculate the areas of the faces and record the results below.

    Areas of the lateral faces: _____

    Areas of the bases: _____

    d. Add the areas of the lateral faces to find the lateral surface area of the prism.

    _____

    e. Add the areas of all the faces to find the total surface area of the prism.

    _____

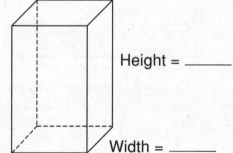

Height = _____

Width = _____

Length = _____

2. a. Find a model of a pyramid with a square base in the set of geometric solids.

    b. Measure the length of the base and the slant height of the pyramid. Record the measurements in the figure.

    c. Calculate the areas of the faces and record the results.

    Areas of the lateral faces:

    _____

    Area of the base:

    _____

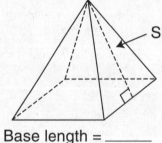

Slant height = _____

Base length = _____

Holt Mathematics

TEKS 8.8.A

★ LAB
8-8B # Using Models to Find Lateral and Total Surface Areas

d. Add the areas of the lateral faces to find the lateral surface area of the pyramid.

_____

e. Add the areas of all the faces to find the total surface area of the pyramid.

_____

3. a. Find a model of a cylinder in the set of geometric solids.

b. Measure the diameter and height of the cylinder. Record the measurements in the figure.

c. Calculate the areas of the surfaces and record the results.

Area of the lateral surface: _____

Areas of the bases: _____

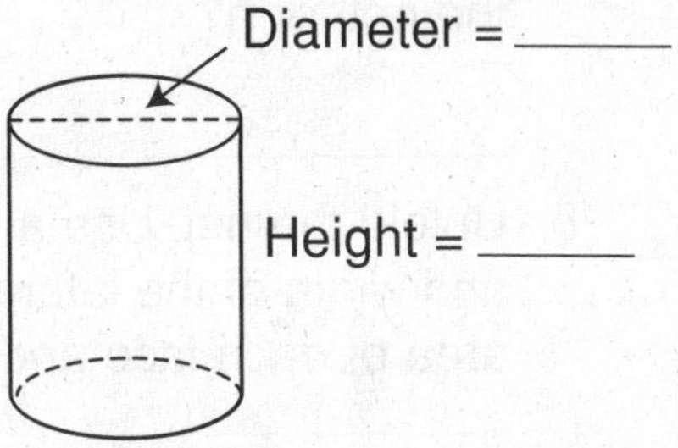

Diameter = _____

Height = _____

d. What is the lateral surface area of the cylinder?

_____

e. Add the areas of all the surfaces to find the total surface area of the cylinder.

_____

Try This

1. **Write About It**   Choose a different three-dimensional figure from the set of solids and explain the steps in finding the total surface area of the figure.

_____

_____

_____

_____

Holt Mathematics

Name _____ Date _____ Class _____

TEKS 8.8.A

LAB 8-8C # Using Nets to Find Lateral and Total Surface Areas

Use with Lesson 8-8

Materials needed: pencil, scissors, inch ruler
Activity
You can use nets to help you find lateral and total surface areas of three-dimensional figures.

1. **a.** Cut out and fold Net #1. What three-dimensional figure does the net form?

_____

b. Unfold the net. Use a ruler to measure, in inches, the length and width of the lateral faces and the bases. Calculate the area of each face and fill in the table.

Face	Area
Lateral Face A	
Lateral Face B	
Lateral Face C	
Lateral Face D	
Base A	
Base B	

c. Add the areas of the lateral faces to find the lateral surface area of the three-dimensional figure.

_____

d. Add the areas of all the faces to find the total surface area of the three-dimensional figure.

_____

Holt Mathematics

TEKS 8.8.A

  Using Nets to Find Lateral and Total Surface Areas

Net #1

Base B

Lateral Face D

Lateral Face C

Lateral Face B

Lateral Face A

Base A

Name _____ Date _____ Class _____

Using Nets to Find Lateral and Total Surface Areas
LAB 8-8C

2. a. Cut out and fold Net #2. What three-dimensional figure is formed?

_____

b. Unfold the net. Use a ruler to measure the base and the lateral faces. Use your measurements to calculate the area of each face and fill in the table.

Face	Area
Lateral Face A	
Lateral Face B	
Lateral Face C	
Lateral Face D	
Base	

c. Add the areas of the lateral faces to find the lateral surface area of the three-dimensional figure.

_____

d. Add the areas of all the faces to find the total surface area of the three-dimensional figure.

_____

Holt Mathematics

TEKS 8.8.A

Using Nets to Find Lateral and Total Surface Areas

Net #2

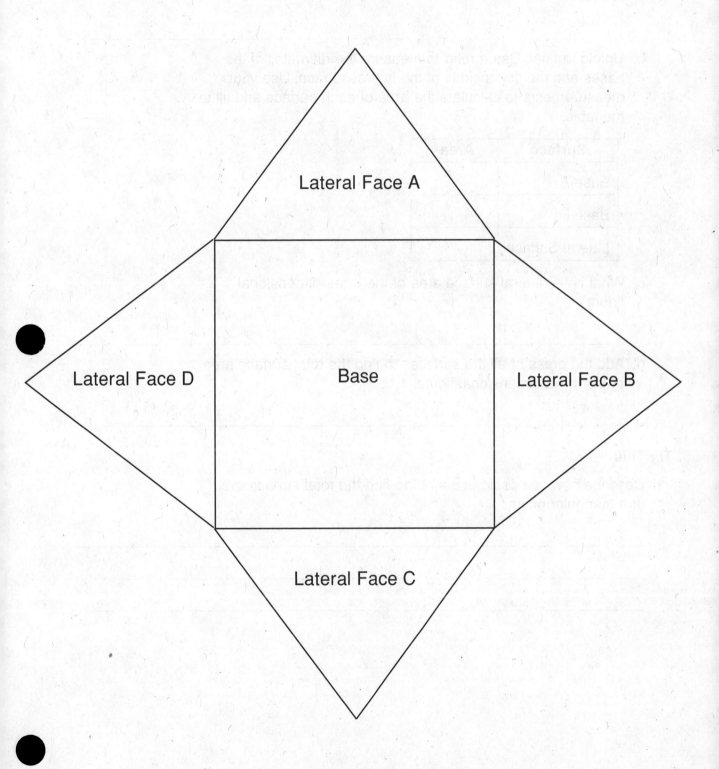

TEKS 8.8.A

LAB
8-8C
Using Nets to Find Lateral and Total Surface Areas

3. a. Cut out and fold Net #3. What three-dimensional figure is formed?

_____

b. Unfold the net. Use a ruler to measure the diameter of the bases and the dimensions of the lateral surface. Use your measurements to calculate the area of each surface and fill in the table.

Surface	Area
Base A	
Base B	
Lateral Surface	

c. What is the lateral surface area of the three-dimensional figure?

_____

d. Add the areas of all the surfaces to find the total surface area of the three-dimensional figure.

_____

Try This

1. Describe how you could use a net to find the total surface area of a triangular prism.

_____

_____

_____

Holt Mathematics

TEKS 8.8.A

  Using Nets to Find Lateral and Total Surface Areas

LAB 8-8C

Net #3

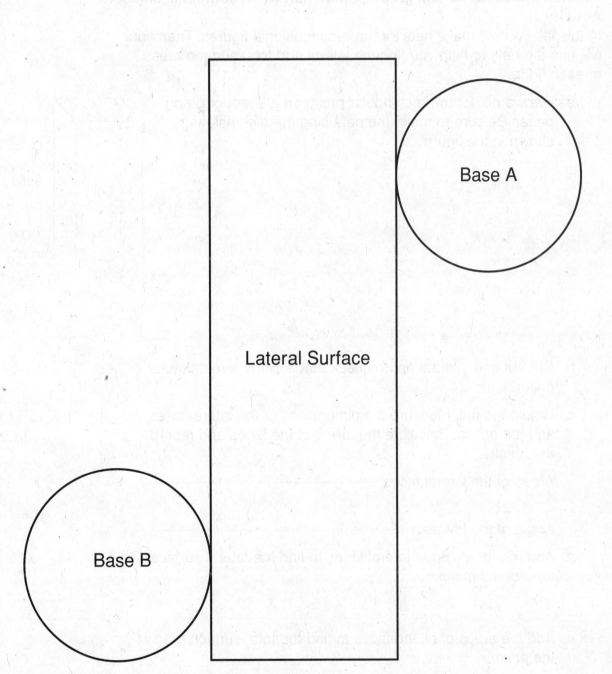

Base A

Lateral Surface

Base B

Holt Mathematics

Name _____ Date _____ Class _____

 LAB 8-8D **Drawing Nets to Find Lateral and Total Surface Areas**

Use with Lesson 8-8

Materials needed: pencil, graph paper, inch ruler, compass, scissors
Activity

In this lab you will make nets for three-dimensional figures. Then you will use the nets to help you find the lateral and total surface areas of each figure.

1. a. Draw a net for the rectangular prism on a sheet of graph paper. Be sure to make the net using the dimensions shown in the figure.

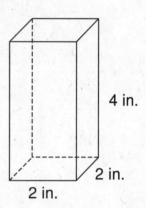

4 in.

2 in.

2 in.

  b. Cut out and fold the net to check that it forms a rectangular prism.

  c. Unfold the net. Measure the dimensions of the lateral faces and the bases. Calculate the areas of the faces and record the results.

   Areas of the lateral faces: ———————————————

   Areas of the bases: ———————————————

  d. Add the areas of the lateral faces to find the lateral surface area of the prism.

———————————————————————————————————

  e. Add the areas of all the faces to find the total surface area of the prism.

———————————————————————————————————

 Holt Mathematics

TEKS 8.8.A

  Drawing Nets to Find Lateral and Total Surface Areas

2. a. Draw a net for the pyramid on a sheet of graph paper using the dimensions shown in the figure.

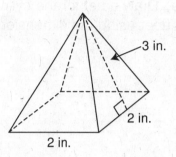

3 in.

2 in.

2 in.

b. Cut out and fold the net to check that it forms a pyramid.

c. Unfold the net. Measure the dimensions of the lateral faces and the base. Calculate the areas of the faces and record the results.

Areas of the lateral faces: _____

Area of the base: _____

d. Add the areas of the lateral faces to find the lateral surface area of the pyramid.

_____

e. Add the areas of all the faces to find the total surface area of the pyramid.

_____

Holt Mathematics

TEKS 8.8.A

LAB
8-8D **Drawing Nets to Find Lateral and Total Surface Areas**

3. a. Draw a net for the cylinder on a sheet of graph paper using the dimensions shown in the figure.

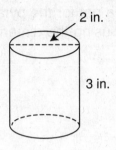

2 in.

3 in.

   b. Cut out and fold the net to check that it forms a cylinder.

   c. Unfold the net. Measure the dimensions of the lateral surface and the bases. Calculate their areas and record the results.

      Area of the lateral surface: _____

      Areas of the bases: _____

   d. What is the lateral surface area of the cylinder?

      _____

   e. Add the areas of all the surfaces to find the total surface area of the cylinder.

      _____

Try This

1. **Critical Thinking**   Describe a three-dimensional figure whose net is made up of squares. What shortcut can you use to find the total surface area of the figure?

   _____

   _____

Holt Mathematics

Name _____ Date _____ Class _____

TEKS 8.8.B

  Using Models to Explore the Volume of a Sphere

Use with Lesson 8-9

Materials needed: pencil; a set of Relational GeoSolids; sand, rice, or other material that can be used to fill the solids

Activity

In this lab you will use models to explore relationships among cylinders, cones, and spheres. You will use what you discover to write a formula for the volume of a sphere.

1. Find GeoSolids for a cone, sphere, and cylinder that all have the same radius and the same height.

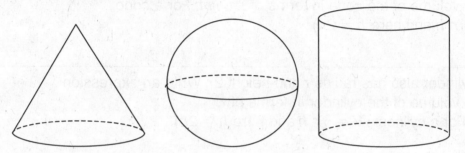

2. Fill the cone with sand or other material. Carefully pour the sand from the cone into the sphere. Then refill the cone and pour the sand into the sphere again. Repeat the process as needed until you have filled the sphere. How is the volume of the sphere related to the volume of the cone?

_____

3. Fill the cone with sand. Carefully pour the sand from the cone into the cylinder. Then refill the cone and pour the sand into the cylinder again. Repeat the process as needed until you have filled the cylinder. How is the volume of the cylinder related to the volume of the cone?

_____

4. Suppose the cone has a volume of 1 in.^{3} What is the volume of the sphere? What is the volume of the cylinder?

_____

5. Explain why the volume of the sphere is the average of the volume of the cone and the volume of the cylinder.

_____

Holt Mathematics

Using Models to Explore the Volume of a Sphere

6. Let r be the radius of the sphere.

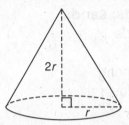 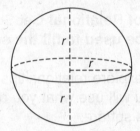 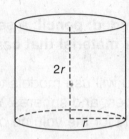

Then the cone has radius r and height $2r$. Write an expression for the volume of the cone in terms of r. (*Hint:* For a cone, $V = \frac{1}{3}\pi r^2 h$ and here $h = 2r$.)

_____

The cylinder also has radius r and height $2r$. Write an expression for the volume of the cylinder in terms of r.
(*Hint:* For a cylinder, $V = \pi r^2 h$ and here $h = 2r$.)

_____

7. The volume of the sphere is the average of the expressions for the volume of the cone and the volume of the cylinder. Write the expressions for the volume of the cone and the volume of the cylinder from Step 6. Then simplify to find a formula for the volume of a sphere.

$V = \frac{1}{2}$ (_____ + _____) = _____

Try This

Use the formula you discovered to find the volume of a sphere with the given radius. Use 3.14 for π and round your answers to the nearest tenth.

1. $r = 2$ ft _____   **2.** $r = 6$ cm _____

Holt Mathematics

Name _____ Date _____ Class _____

TEKS 8.12.C

Using Graphs to Display Data About Presidents

Use with Lesson 9-5

Materials needed: pencil, software for making graphs, computer

Activity

In this lab you will use technology to make graphs that display data about U.S. presidents. The table shows the age at inauguration and the party (Republican or Democrat) of all the presidents who served during the 20th century.

20th-Century Presidents					
Age	**Party**	**Age**	**Party**	**Age**	**Party**
54	Rep.	54	Rep.	56	Rep.
42	Rep.	51	Dem.	61	Rep.
51	Rep.	60	Dem.	52	Dem.
56	Dem.	62	Rep.	69	Rep.
55	Rep.	43	Dem.	64	Rep.
51	Rep.	55	Dem.	46	Dem.

1. Display the data about the ages of the presidents by using technology to make a line plot, a stem-and-leaf plot, and a box-and-whisker plot. For each graph, tell which type of technology you used.

_____

_____

Holt Mathematics

Using Graphs to Display Data About Presidents

2. Use technology to make a circle graph that displays the percentage of presidents of each party. Tell which type of technology you used.

_____

3. Use technology to make a Venn diagram that shows how many Republican presidents were over the age of 53 at their inauguration. Tell which type of technology you used.

_____

Holt Mathematics

TEKS 8.12.C

 LAB 9-5A

Using Graphs to Display Data About Presidents

Try This

1. Which of the graphs that you made would be most useful for someone who wanted detailed information about the presidents' ages? Why?

_____

_____

2. Which of the graphs that you made would be most useful for someone who wanted to know the median age of the presidents? Why?

_____

_____

_____

_____

Holt Mathematics

TEKS 8.12.C

Using Graphs to Display Nutrition Data

Use with Lesson 9-5

Materials needed: pencil, software for making graphs, computer
Activity
In this lab you will use technology to make graphs that display
nutrition data. The table gives the number of calories, the number of
grams of fiber, and the number of grams of carbohydrates in a one-
cup serving of ten different breakfast cereals.

Nutritional Information for Breakfast Cereals					
Calories	Fiber (g)	Carbohydrates (g)	Calories	Fiber (g)	Carbohydrates (g)
143	1	31	134	5	32
150	3	29	131	1	30
135	5	34	146	1	22
151	1	36	146	2	35
151	1	36	134	2	30

1. Make a line plot showing the number of grams of fiber in the
   cereals. Tell which type of technology you used to make the
   graph.

   _____

   _____

Holt Mathematics

TEKS 8.12.C

Using Graphs to Display Nutrition Data

2. Make a box-and-whisker plot showing the number of grams of carbohydrates in the cereals. Tell which type of technology you used to make the graph.

_____

3. Make a stem-and-leaf plot showing the number of calories in the cereals. Tell which type of technology you used to make the graph.

_____

4. Make a Venn diagram that shows how many of the cereals have greater than 140 calories but less than 35 grams of carbohydrates. Tell which type of technology you used to make the graph.

_____

Holt Mathematics

Using Graphs to Display Nutrition Data

5. Make a circle graph showing the percentage of the cereals with less than 140 calories, the percentage with 140 to 149 calories, and the percentage with greater than 149 calories. Tell which type of technology you used.

_____

Try This

1. Which of the graphs that you made could be used to find the mode of the data that is displayed? Explain.

_____

_____

Holt Mathematics

Name _____ Date _____ Class _____

TEKS 8.12.C

  LAB **9-7** ## Collecting Data to Make a Graph

Use with Lesson 9-7

Materials needed: pencil, centimeter ruler, six square objects of different sizes (for example: tiles, photographs, tangram pieces, sticky notes, etc.)

Activity

In this lab you will make measurements, organize your data, and display the data in a graph. Then you will use your graph to make predictions.

1. Gather six square objects. For each object, first measure the side length to the nearest tenth of a centimeter. Then measure the diagonal length to the nearest tenth of a centimeter. Be sure to measure as accurately as possible. Record the measurements in the table.

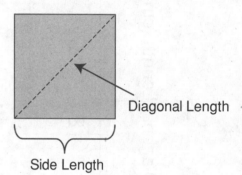

Diagonal Length

Side Length

	Side Length	**Diagonal Length**
Object 1		
Object 2		
Object 3		
Object 4		
Object 5		
Object 6		

2. Which type of graph is best for displaying the data you collected? Explain.

_____

_____

_____

_____

Holt Mathematics

TEKS 8.12.C

**LAB
9-7**

Collecting Data to Make a Graph

3. For each object, use the data in your table to plot an ordered
   pair on the coordinate plane.

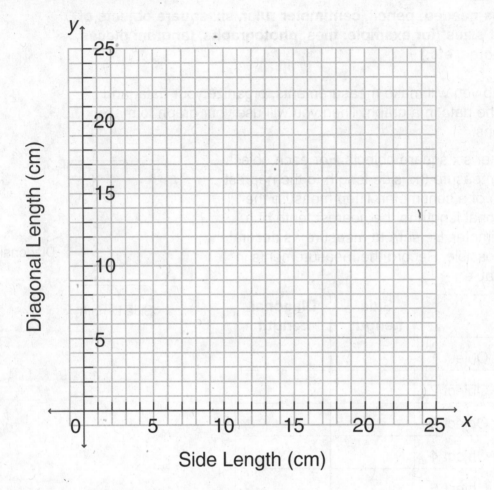

4. What do you notice about the points you plotted?

_____

5. Connect the points you plotted. What type of graph have you
   created?

_____

Holt Mathematics

TEKS 8.12.C

Collecting Data to Make a Graph

Try This

1. Use your graph to predict the diagonal length of a square object that has a side length of 6 cm.

   _____

2. Use your graph to predict the side length of a square object that has a diagonal length of 15 cm.

   _____

3. **Critical Thinking**   What would happen if you collect data by measuring the side length and diagonal length of rectangular objects that are not square?

   _____

Holt Mathematics

Name _____ Date _____ Class _____

Selecting Graphs to Display Internet Data

Use with Lesson 9-8

Materials needed: pencil, software for making graphs, computer

Activity

In this lab you will select the type of graph that is best for displaying
a set of data and then you will use technology to make the graph.
You may choose from any of the following types of graphs.

Line plot	Line graph	Bar graph	Histogram
Stem-and-leaf plot	Circle graph	Venn diagram	Box-and-whisker plot

1. A 2005 survey identified the ages of people
who had used the Internet in the previous 30
days. The table shows the results.

Which type of graph is best for displaying this
data?

_____

Use technology to make the graph.

Ages of Internet Users	
Age Range	**Percent of All Users**
18 to 34	37%
35 to 54	44%
55 and up	19%

2. The table shows the number of Internet users
in various countries in 2004.

Which type of graph is best for displaying this
data?

_____

Use technology to make the graph.

3. The Internet has grown rapidly since the 1990s.
The table gives the number of host Web sites
from 1998 to 2005.

Internet Use in 2004	
Country	**Number of Users (millions)**
Canada	20
France	25
Germany	42
Brazil	22
India	37
Japan	78

Growth of the Internet								
Year	**1998**	**1999**	**2000**	**2001**	**2002**	**2003**	**2004**	**2005**
Number of Hosts (millions)	30	43	72	110	147	172	233	318

Which type of graph is best for displaying this data?

_____

Use technology to make the graph.

Holt Mathematics

TEKS 8.12.C

Selecting Graphs to Display Internet Data

4. A 2002 survey asked Americans where they used the Internet. The results are shown in the table.

Which type of graph is best for displaying this data?

_____

Use technology to make the graph.

Where Do You Use the Internet?	
Location	**Number of Responses**
Home only	36
Home and work	34
Work only	7

5. The following data gives the number of unique visitors (in millions) to ten different travel Web sites during a one-month period in 2005.

9 13 9 6 8 8 9 23 21 14

Name four types of graphs that could be used for displaying this data.

_____

Use technology to make the graphs.

Try This

1. Which of the graphs that you made could be used to make a prediction about the future? Explain.

_____

_____

2. Which of the graphs that you made shows how sets of data overlap? Explain.

_____

_____

Holt Mathematics

Name _____ Date _____ Class _____

Selecting Representations for Data About Texas

Use with Lesson 9-8

Materials needed: pencil

Activity

A travel organization is preparing information about cities in Texas.
They can display the information using any of the following types of
graphs.

Line plot	Line graph	Bar graph	Histogram
Stem-and-leaf plot	Circle graph	Venn diagram	Box-and-whisker plot

1. The table shows the population growth of Dallas.

Population of Dallas, Texas				
Year	1970	1980	1990	2000
Population (millions)	1.6	2.1	2.7	3.5

Which type of graph should the organization use to display this
data? Justify your choice.

_____

2. The organization surveyed 200
people who visited Texas. The table
shows how many of the visitors went
to San Antonio, Austin, both cities, or
neither city.

Which type of graph should the
organization use to display this data?
Justify your choice.

	Number of Responses
Visited San Antonio only	60
Visited Austin only	55
Visited both cities	25
Visited neither city	60

_____

3. These are the daily high temperatures (in degress Fahrenheit) in
Houston during a three-week period.

81 84 84 79 78 84 81 86 79 87 91 87 80 72 73 85 90 91 92 93 87

Which four types of graphs would be appropriate for displaying
this data?

_____

Holt Mathematics

TEKS 8.12.C

LAB
9-8B

Selecting Representations for Data About Texas

4. The table shows the population of several Texas cities according to the 2000 census.

   Which type of graph should the organization use to display this data? Justify your choice.

   _____

   _____

City	Population (thousands)
El Paso	564
Plano	222
Fort Worth	535
Austin	657
Amarillo	174
Waco	113

5. The table shows the ages of people living in Texas according to the U.S. Census Bureau.

   Which type of graph should the organization use to display this data? Justify your choice.

   _____

   _____

Age Group	Percent
Under 18 years	28%
19 to 64 years	62%
65 years or older	10%

Try This

1. Give an example of data about Texas that can best be displayed with a bar graph. Explain why a bar graph would be the best choice to display the data.

   _____

   _____

2. Give an example of data about Texas that can best be displayed with a line graph. Explain why a line graph would be the best choice to display the data.

   _____

   _____

   _____

   _____

Holt Mathematics

Name _____ Date _____ Class _____

Using Graphs to Display Basketball Data

Use with Lesson 9-8

Materials needed: pencil, software for making graphs, computer
Activity
In this lab you will use technology to make graphs that display data
about the WNBA's Houston Comets. The table shows the heights
and ages of the players on the 2005 team.

Height (in.)	71	69	71	66	76	74	77	66	72	74
Age (yr)	36	28	23	26	25	36	25	35	34	30

1. Use technology to display the data about the player's heights in
a line plot, a stem-and-leaf plot, and a box-and-whisker plot. For
each graph, tell which type of technology you used.

_____

_____

_____

Holt Mathematics

TEKS 8.12.C

  Using Graphs to Display Basketball Data

LAB
9-8C

2. Use technology to make a circle graph that displays the percentage of players who are more than 6 feet tall (72 in.) and the percentage of players who are 6 feet tall or less. Tell which type of technology you used.

_____

_____

3. Use technology to make a Venn diagram that shows how many players who are more than 6 feet tall are under the age of 30. Tell which type of technology you used.

_____

_____

Try This

1. Give an example of a different problem that you could solve using the given data and a Venn diagram.

_____

_____

Holt Mathematics

TEKS 8.12.C

LAB
9-8C

Using Graphs to Display Basketball Data

2. a. Write a problem that can be solved using the line plot that
you made.

_____

b. Would it also be possible to solve this problem using the box-
and-whisker plot? Why or why not?

_____

_____

_____

Holt Mathematics

Name _____ Date _____ Class _____

TEKS 8.12.C

Selecting Representations to Display Movie Data

Use with Lesson 9-2

Materials needed: pencil, software for making graphs, computer

Activity

In this lab, you will use technology to display data about movies. You may use any of the following types of graphs.

Line plot	Line graph	Bar graph	Histogram
Stem-and-leaf plot	Circle graph	Venn diagram	Box-and-whisker plot

1. The table shows the amount of money made in the United States by the top five films of all time.

Select the best type of graph to display this data. Then use technology to make the graph. Tell which type of technology you used.

Box-Office Champions	
Movie	**Revenue (millions of dollars)**
Titanic	601
Star Wars	461
Shrek 2	441
E.T.	435
Star Wars: Episode I	431

Holt Mathematics

Selecting Representations to Display Movie Data

LAB 9-8D

2. The manager of a movie theater took a survey of 30 moviegoers. He asked them what types of films they had seen in the past month. The table shows the results.

Select the best type of graph to display this data. Then use technology to make the graph. Tell which type of technology you used.

_____

Survey Results	
Films Seen in the Past Month	**Number of Moviegoers**
Action only	11
Comedy only	5
Drama only	3
Action and comedy	2
Comedy and drama	3
Drama and action	1
All three types	5

Holt Mathematics

Name _____ Date _____ Class _____

  Selecting Representations to Display Movie Data

3. The table shows the average price of a movie ticket in various years.

Ticket Prices						
Year	1980	1985	1990	1995	2000	2005
Average Ticket Price ($)	2.69	3.55	4.23	4.35	5.39	6.40

Select the best type of graph to display this data. Then use technology to make the graph. Tell which type of technology you used.

_____

_____

Holt Mathematics

Name _____ Date _____ Class _____

LAB 9-8D **Selecting Representations to Display Movie Data**

4. The following data shows the amount of money (in millions of dollars) that was earned by 12 films in 2005.

51 42 33 32 33 47 42 32 32 39 46 33

Select the four best types of graphs to display this data. Then use technology to make the graphs. Tell which type of technology you used for each graph.

_____

_____

_____

_____

Holt Mathematics

TEKS 8.12.C

⬟ LAB 9-8D Selecting Representations to Display Movie Data

5. The 1997 film *Titanic* made 33% of its revenue in the United States. The remaining 67% of its revenue came from overseas.

   Select the best type of graph to display this data. Then use technology to make the graph. Tell which type of technology you used.

   _____

Try This

1. Write a problem that can be solved using one of the graphs that you made. Include the answer to your problem.

   _____

   _____

2. Which of the graphs that you made shows how data changes over time? Describe how the data changes.

   _____

   _____

   _____

   _____

Holt Mathematics

Name _____ Date _____ Class _____

TEKS 8.12.C

Selecting Representations for Super Bowl Data

Use with Lesson 9-2

Materials needed: pencil
Activity
Sports historians are preparing a presentation about the Super Bowl. They can display their data using any of the following types of graphs.

Line plot	Line graph	Bar graph	Histogram
Stem-and-leaf plot	Circle graph	Venn diagram	Box-and-whisker plot

1. These are the scores of the winning team in each of the Super Bowls from 1987 to 2006.

   39 42 20 55 20 37 52 30 49 27 35 31 34 23 34 20 48 32 24 21

   Which four types of graphs would be best for displaying this data?

   _____

2. From 1997 to 2006, teams from the AFC won 70% of the Super Bowl games and teams from the NFC won 30% of the games.

   Which type of graph should the historians use to display this data? Justify your choice.

   _____

3. In the 2006 Super Bowl, the Pittsburgh Steelers played the Seattle Seahawks. The historians surveyed 100 people who watched the game to see which team they preferred.

   Which type of graph should the historians use to display this data? Justify your choice.

	Number of Responses
Steelers fan	50
Seahawks fan	29
Fan of both teams	21

_____

_____

Holt Mathematics

TEKS 8.12.C

Selecting Representations for Super Bowl Data

4. The table shows the average cost of a 30-second commercial during the Super Bowl.

Which type of graph should the historians use to display this data? Justify your choice.

_____

_____

Cost of a 30-Second Commercial During the Super Bowl	
Year	Cost (millions of dollars)
1996	1.1
1997	1.2
1998	1.3
1999	1.6
2000	2.1
2001	2.1
2002	1.9
2003	2.1
2004	2.2
2005	2.4

5. The table shows the number of times that several teams played in the Super Bowl from 1997 to 2006.

Which type of graph should the historians use to display this data? Justify your choice.

_____

_____

_____

Team	Number of Times in Super Bowl (1997-2006)
New England Patriots	4
Denver Broncos	2
Pittsburgh Steelers	1
Green Bay Packers	2
Atlanta Falcons	1

Holt Mathematics

Name _____ Date _____ Class _____

TEKS 8.12.C

Selecting Representations for Super Bowl Data

Try This

1. Describe an example of sports data that can best be displayed
   with a line graph. Explain why a line graph is the best choice for
   displaying the data.

   _____

   _____

   _____

   _____

Holt Mathematics

TEKS 8.12.C

Selecting Representations for Data About Summer

LAB 9-8F

Use with Lesson 9-8

Materials needed: pencil
Activity
Writers for the school newspaper are preparing a feature article about summer vacation. They can display the information using any of the following types of graphs.

Line plot	Line graph	Bar graph	Histogram
Stem-and-leaf plot	Circle graph	Venn diagram	Box-and-whisker plot

1. One reporter surveyed 50 students about whether they attended summer camp, went on a family vacation, or stayed in town. The results of the survey are shown in the table. Which type of graph should the reporter use to display this data? Justify your choice.

Summer Trips	Number of Students
Attended summer camp	18
Went on family vacation	29
Attended summer camp and went on family vacation	7
Stayed in town	10

_____

_____

2. Another reporter asked 60 students what their favorite summer activities were. The results of the survey are shown in the table. Which type of graph should the reporter use to display this data? Justify your choice.

Favorite Activity	Percent
Swimming	45%
Playing Sports	30%
Arts and Crafts	10%
Traveling	10%
Other	5%

_____

Try This

1. Give an example of data about summer that can best be displayed with a Venn diagram. Explain why a Venn diagram would be the best choice to display this data.

_____

_____

Holt Mathematics

Name _____ Date _____ Class _____

LAB 10-3 **Use a Simulation**

Use with Lesson 10-3

Materials needed: computer or calculator, pencil
Activity
Use a computer or calculator to create a table of random numbers.
Then use the table to record at least ten trials to simulate the
following probability.

Lisa's school bus is late about 25% of the time. What is the
probability that her bus will be late at least 3 days in a 5-day school
week?

	Mon	Tue	Wed	Thur	Fri	Times bus is late
Trial 1						
Trial 2						
Trial 3						
Trial 4						
Trial 5						
Trial 6						
Trial 7						
Trial 8						
Trial 9						
Trial 10						

Step 1 Identify the important information given in the problem.

_____

_____

Step 2 Use a table to list randomly generated numbers from 00 to
99 for ten trials. Since you want to find the probability that Lisa's bus
will be late in the next 5 days, 5 numbers will make up a trial.

Holt Mathematics

TEKS 8.11.C

★ LAB 10-3

Use a Simulation

Step 3 Determine how you will gather the data.

Which numbers in the table will represent Lisa's bus being late?

Which numbers in the table will represent Lisa's bus being on time?

_____

_____

Step 4 Count the number of times Lisa's bus is late in each trial. Next count the number of trials in which Lisa's bus is late at least 3 times.

Divide this number by 10 (the number of trials) to find the experimental probability that Lisa's bus is late at least 3 days in a 5-day week.

Step 5 Based on the fact that Lisa's bus is late only about 25% of the time, is your answer reasonable? Explain.

_____

_____

_____

Think and Discuss

1. How did you know what numbers to choose for Lisa's bus to be on time and for Lisa's bus to be late?

_____

_____

_____

_____

2. Jana claims that her simulation should have resulted in the bus being late 60% of the time. Identify and correct her error.

_____

_____

_____

Holt Mathematics

TEKS 8.11.C

⭐ **LAB 10-3** **Use a Simulation**

Try This

Create a table of random numbers to simulate the following situations. Use at least 10 trials for each simulation. Follow the steps in the Activity section.

1. Repeat the simulation in the Activity section using 30% as the probability that Lisa's bus will be late. According to the simulation, what is the experimental probability that it will be late? Does this result support the usefulness of random number simulations?

   _____

   _____

   _____

   _____

2. 18 of the 20 students in a math class passed a quiz. Perform a simulation to find the experimental probability that one student randomly chosen from the class did not pass the quiz.

   _____

   _____

3. The probability that you will get a 4 when you toss a number cube is $\frac{1}{6}$. Perform a simulation to find the experimental probability you will get a 4 or 3 out of the next 10 tosses.

   _____

   _____

Holt Mathematics

Name _____ Date _____ Class _____

TEKS 8.4, 8.7.D

LAB
12-1
Graphing Linear Equations

Use with Lesson 12-1

If all of the variables in an equation are raised only to the first power, the equation is known as a linear equation. The equation $y = 3x + 7$ is a linear equation because both x and y are raised to the first power.

Activity

1. Complete the following table.

x	1	2	3	4	5	6	8	10
$y = 3x + 7$	10							

2. Predict what the graph of the data you have generated would look like.

_____

3. Graph the data from part 1.
   The graph matches your prediction from part 2.

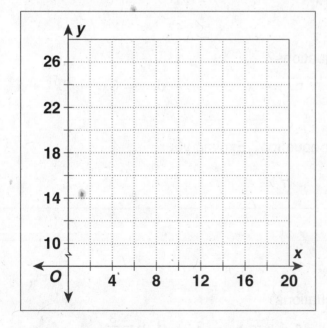

Holt Mathematics

TEKS 8.4, 8.7.D

Graphing Linear Equations

4. Complete the following table and draw a graph as in part 1.

x	1	2	3	4	5
$y = x^2$					

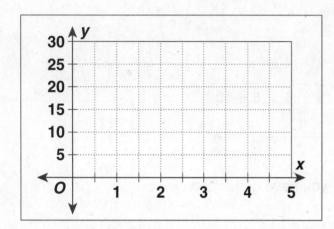

Explain why linear equations are called linear.

_____

Think and Discuss

1. Which of the following are linear equations?

a. $y = 3x$ **b.** $x + 6 = y - 3$ **c.** $y = x^2$ **d.** $y = 3x^3 + 2x^2 + x$

_____   _____   _____   _____

2. Is the equation $y = \sqrt{x + 2}$ a linear equation? Explain your answer.

_____

_____

Try This

1. Which of the following are linear equations?

a. $y = 2x - 7$ **b.** $x = y - 8$ **c.** $3y = x^2 + 3$ **d.** $y = \dfrac{3}{x}$

_____   _____   _____   _____

Holt Mathematics

TEKS 8.3.A

Direct Variation

Use with Lesson 12-5

Activity 1

Evie is deciding between a monthly phone plan and a pay as you go rate. Plan A charges $0.15 for each minute. Plan B charges $12 per month to maintain the phone line and $0.10 for each minute on the phone.

Number of minutes	50 min	100 min	150 min	200 min	250 min	300 min
Plan A	$ 7.50	$15.00	$22.50	$30.00	$37.50	$45.00
Plan B	$17.00	$22.00	$27.00	$32.00	$37.00	$42.00

Step 1 Make a graph comparing the two phone plans based on the number of minutes used per month. Use a different color for each phone plan.

Remember: A direct variation between the number of minutes on the phone plan needs to satisfy two conditions: the relationship is linear and there is a constant ratio to describe that relationship.

Step 2 Compare the number of minutes on the phone to the cost of Plan A and to the cost of Plan B to determine if each relationship is linear.

_____

_____

_____

_____

Holt Mathematics

Name _____ Date _____ Class _____

LAB 12-5A

Direct Variation

Step 3 Compare ratios between the number of minutes used and cost for both plans. Are the ratios constant? If they are, what is the constant ratio?

Plan A: $\frac{7.50}{50}$ = _____ , $\frac{15}{100}$ = _____ , $\frac{22.50}{150}$ = _____ ,

$\frac{30.00}{200}$ = _____ , $\frac{37.50}{250}$ = _____ , $\frac{45.00}{300}$ = _____ ,

_____

Plan B: $\frac{17}{50}$ = _____ , $\frac{22}{100}$ = _____ , $\frac{27}{150}$ = _____ . . . .

_____

Does Plan A show a direct variation? _____

Does Plan B show a direct variation? _____

Think and Discuss

1. How would you use the data in the table to find an equation of direct variation for one of the phone plans? Show your work.

_____

_____

_____

Holt Mathematics

Name _____ Date _____ Class _____

TEKS 8.3.A

Direct Variation

LAB 12-5A

Try This

1. Graph the data or compare ratios to determine whether the data set shows direct variation.

Number of Hours Studied	1	2	3	4	5
Test Scores	92	84	88	85	91

Does the data set show direct variation? _____

2. Does the data in the graph show direct variation? _____

If so, what is the common ratio? _____

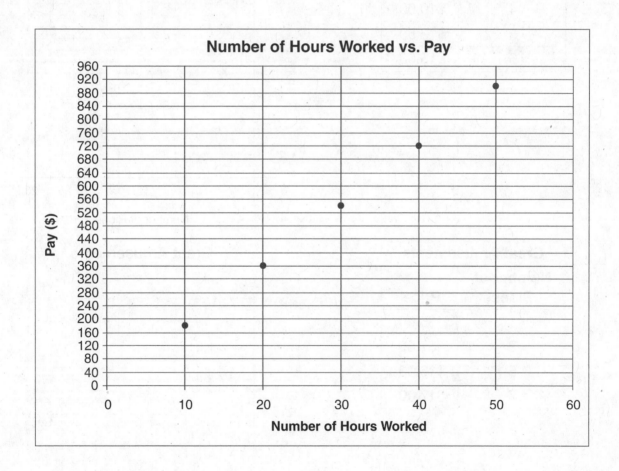

Holt Mathematics

Name _____ Date _____ Class _____

TEKS 8.3.A

 Proportional and Nonproportional
12-5B Relationships

Use with Lesson 12-5

Materials needed: pencil
Activity

1. The tables show the price of rides at four different amusement
   parks. For each park, make a graph that shows the relationship
   between the number of rides and the price.

a.

Fun Zone	
Number of Rides	Price
2	$5.00
4	$10.00
5	$12.50
6	$15.00

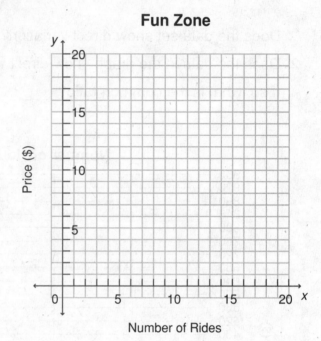

b.

Great Amusements	
Number of Rides	Price
1	$5.00
2	$7.00
5	$13.00
7	$17.00

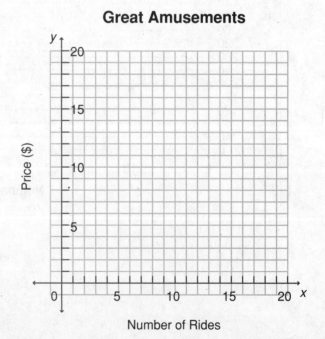

Holt Mathematics

TEKS 8.3.A

Proportional and Nonproportional Relationships

LAB 12-5B

c.

Adventure Town	
Number of Rides	Price
1	$3.00
3	$9.00
4	$12.00
6	$18.00

Adventure Town

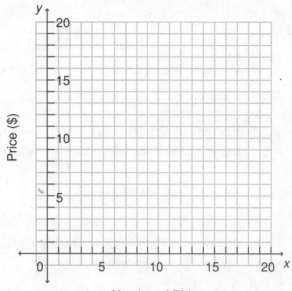

d.

Action Park	
Number of Rides	Price
2	$7.00
3	$8.50
6	$13.00
8	$16.00

Action Park

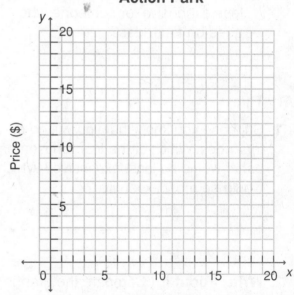

2. a. For which amusement parks is there a direct variation between the number of rides and the price? How can you tell?

_____

_____

Holt Mathematics

TEKS 8.3.A

Proportional and Nonproportional
LAB 12-5B Relationships

b. For each of these parks, write an equation for the direct variation.

_____

c. What is the price per ride at each of these parks?

_____

3. a. For which amusement parks is there NOT a direct variation between the number of rides and the price? Why?

_____

_____

b. Graph the data sets for these parks on the same coordinate plane.

c. If you haven't already done so, draw a line through the data points for each of these parks. Which park is less expensive for someone who goes on four rides?

_____

Try This

1. Suppose you know that you will go on four rides. Which of the four amusement parks would be least expensive? Which would be most expensive?

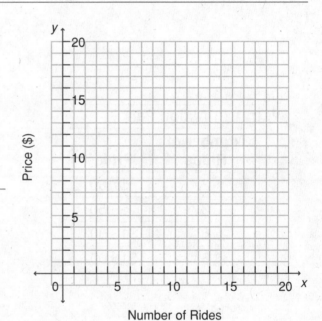

_____

_____

2. Write About It Compare the graphs for the parks where there is a direct variation relationship to the graphs for the parks where there is not a direct variation relationship. How are the graphs similar? How are they different?

_____

_____

_____

Holt Mathematics

Name _____ Date _____ Class _____

Terms of Geometric Sequences

LAB 13-2

Use with Lesson 13-2

Activity 1

Find the common ratio of the geometric sequence and complete the table.

Term Number	a_1	a_2	a_3	a_4	a_5	a_6	a_7	a_8
Term	1	4	16	64				

To find the common ratio, divide each term by the previous term.

1. Divide the second term, a_2, by the first term, a_1.

$4 \div 1 =$ _____

2. Divide the third term, a_3, by the second term, a_2.

$16 \div 4 =$ _____

3. Repeat for the ratios between a_3 and a_4.

$64 \div 16 =$ _____

4. What is the common ratio? _____

To complete the table, multiply each term by the common ratio to find the next term.

5. Multiply a_4 by the common ratio to get a_5.

6. Multiply a_5 by the common ratio to get a_6.

7. Repeat the multiplication to get a_7 and a_8.

Activity 2

Find the 12th term in the geometric sequence: 1, 3, 9, 27 . . .

Step 1 Find the common ratio, r, of the sequence.

$r = 3 \div 1 =$ _____

Step 2 Use the formula $a_n = a_1 r^{n-1}$ to find the 12th term.

$a_{12} = 1(3)^{12-1}$

$a_{12} =$ _____

Holt Mathematics

Name _____ Date _____ Class _____

TEKS 8.16.A

LAB 13-2

Terms of Geometric Sequences

Think and Discuss

1. How would you show that the following sequence is geometric?

512, 128, 32, 8, . . .

_____

_____

2. Complete the following table for a geometric sequence and for an arithmetic sequence. Compare the two sequences.

Term Number	a_1	a_2	a_3	a_4	a_5	a_6	a_7
Term (geometric)	1	4					
Term (arithmetic)	1	4					

_____

_____

Try This

1. Could the following sequence be geometric? If so, what is the common ratio?

2, 5, 8, 11, 14, . . .

_____

_____

2. Find the next 4 terms in the sequence shown in the table.

Term Number	a_1	a_2	a_3	a_4	a_5	a_6	a_7	a_8
Term	2	$\frac{1}{2}$	$\frac{1}{8}$	$\frac{1}{32}$				

3. Find the 18th term of the geometric sequence. Show your work.

160, 80, 40, 20, . . .

Holt Mathematics

1-1 Variables and Expressions

Catherine's dance team is planning a spring trip to the coast. Catherine is saving money in a bank account to pay for the trip. Her parents started her account with $100. She sells Christmas plants and adds $2.50 to her account for each plant she sells.

How much will be in her account if she sells 50 plants?

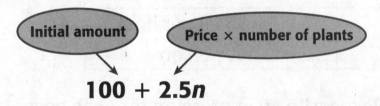

$$100 + 2.5n$$

Evaluate the expression **100 + 2.5n** by substituting 50 for *n*.

$$100 + 2.5(50)$$
$$100 + 125$$
$$225$$

There will be $225.00 in Catherine's account if she sells 50 plants.

Evaluate the expression 100 + 2.5n for each value of *n*.

1. $n = 10$ $\qquad$ $100 + 2.5(10)$

2. $n = 25$ $\qquad$ $100 + 2.5(\quad)$

3. $n = 75$ $\qquad$ $100 + 2.5(\quad)$

Think and Discuss

4. Explain what *n* represents.

5. Describe how you evaluated the expression for different values of *n*.

Holt Mathematics

1-2 Algebraic Expressions

The table shows the prices for different activities at a park.

Activity	Price
Park ride	$1.25 per ride
Jet ski rental	$25.00 + $5.00 per hour
Scooter rental	$10.00 + $2.50 per hour
Bike rental	$5.00 + $2.00 per hour
Bay cruise	$25.00 per person

> An algebraic expression for this cost is 1.25r.

Use a variable to write an expression for each cost.

	Activity	Cost	Variable	Expression
	Park ride	$1.25 per ride	r (rides)	1.25r
1.	Jet ski rental	$25.00 + $5.00 per hour	h (hours)	
2.	Scooter rental	$10.00 + $2.50 per hour	h (hours)	
3.	Bike rental	$5.00 + $2.00 per hour	h (hours)	
4.	Bay cruise	$25.00 per person	p (people)	

To find the cost of renting a jet ski for 5 hours, evaluate the expression $25 + 5h$ by substituting 5 for h.

$$25 + 5(5) = 25 + 25 = 50$$

5. Use your expressions from above to evaluate the cost of a 5-hour scooter rental and of a 5-hour bike rental.

Think and Discuss

6. **Describe** a real-world situation in which you might use the expression $50 + 2x$.

7. **Explain** what you need to write an algebraic expression.

Holt Mathematics

1-3 Integers and Absolute Value

You can use a number line to find the distance of a number from 0. For example, to find the distance of −3 from 0, start at −3 and count the number of units until you get to 0.

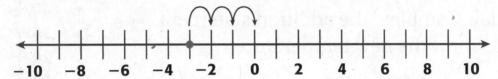

The distance of −3 from 0 is 3.

1. Use a number line to find the distance of each number from 0.
   a. 2
   b. 5
   c. 8
   d. 10
   e. 0

2. Use a number line to find the distance of each number from 0.
   a. −2
   b. −5
   c. −8
   d. −10

Think and Discuss

3. **Describe** any patterns you notice in your results.

4. **Explain** whether it is possible for the distance of a number from 0 to be negative.

Holt Mathematics

1-4 Adding Integers

You can use a thermometer to model addition of integers.

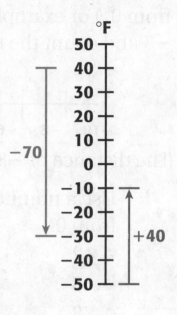

1. Suppose the temperature starts at −50°F and increases 40° during the day. Complete the addition statement to show the new temperature.

$$-50° + 40° = \underline{}$$

2. Suppose the temperature starts at 40°F and drops 70° overnight. Complete the addition statement to show the new temperature.

$$40° + (-70°) = \underline{}$$

Complete the addition statement modeled by each number line.

3. $-4 + \underline{} = 5$

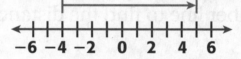

4. $-1 + (\underline{}) = -8$

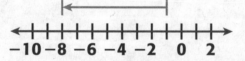

Think and Discuss

5. **Explain** how to add integers on a number line.

Holt Mathematics

1-5 Subtracting Integers

You can use a number line to model subtracting integers.

To subtract 10 from 50, begin at the number being subtracted, 10, and count the number of units to the number 50.

$$50 - 10 = 40$$

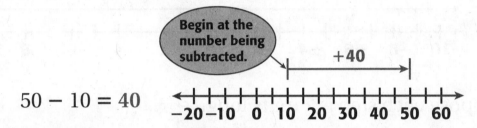

The direction is **right**, so the difference is **positive**.

To subtract -10 from -70, begin at the number being subtracted, -10, and count the number of units to the number -70.

$$-70 - (-10) = -60$$

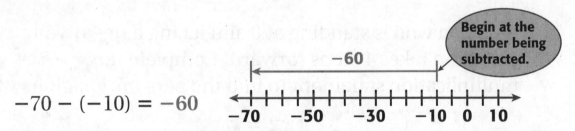

The direction is **left**, so the difference is **negative**.

Use the number line to complete each subtraction statement.

1. $5 - (-4) =$ ___

2. $-8 - (-1) =$ ___

Think and Discuss

3. Discuss a different strategy for subtracting integers.

Holt Mathematics

1-6 Multiplying and Dividing Integers

Imagine a person walking on a number line. If the person faced a **positive direction**, it would be **to the right**. If the person faced a **negative direction**, it would be **to the left**.

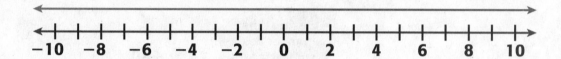

Suppose each step is 2 units long.

1. A person who is standing at 0 and facing a **positive direction** takes 3 steps **backward.** Complete the multiplication statement to find the person's location.

$$3 \cdot (-2) = \underline{\quad}$$

2. A person who is standing at 0 and facing a **negative direction** takes 4 steps **forward.** Complete the multiplication statement to find the person's location.

$$-4 \cdot 2 = \underline{\quad}$$

3. A person who is standing at 0 and facing a **negative direction** takes 5 steps **backward.** Complete the multiplication statement to find the person's location.

$$-5 \cdot (-2) = \underline{\quad}$$

Think and Discuss

4. **Describe** a situation for the multiplication statement $6 \cdot (-3) = -18$ by using a number line.

Holt Mathematics

EXPLORATION

1-7 Solving Equations by Adding or Subtracting

Evaluate the expressions for each given value of *x*.

1.

x	x + 1
0	
1	
2	
3	

2.

x	x − 2
3	
4	
5	
6	

Each expression below has been evaluated. Find the value of *x*.

3.

x	x + 2
	3
	4
	5
	6

4.

x	x − 5
	0
	1
	2
	3

Think and Discuss

5. **Explain** how you evaluated the expressions in Problems 1 and 2.

6. **Explain** how you found the values of *x* in Problems 3 and 4.

Holt Mathematics

1-8 Solving Equations by Multiplying or Dividing

Evaluate the expressions for each given value of *x*.

1.

x	2*x*
0	
1	
2	
3	

2.

x	$\frac{x}{2}$
0	
2	
4	
6	

Each expression below has been evaluated. Find the value of *x*.

3.

x	3*x*
	3
	6
	9
	12

4.

x	$\frac{x}{4}$
	0
	1
	2
	3

Think and Discuss

5. Explain how you evaluated the expressions in Problems 1 and 2.

6. Explain how you found the values of *x* in Problems 3 and 4.

Holt Mathematics

1-9 Introduction to Inequalities

To raise money for a school trip, Angela sells cookies for $2.50 per box. She has to raise at least $100 from the cookie sales in order to go on the trip.

1. Which of the numbers of boxes of cookies shown gives Angela the following amounts of money?

   a. Less than her goal of $100

   b. Exactly $100

   c. More than her goal of $100

25 boxes
30 boxes
40 boxes
50 boxes

2. Name another number of boxes that Angela could sell that would give her less than her goal of $100.

3. Name another number of boxes that Angela could sell that would give her more than her goal of $100.

Think and Discuss

4. **Describe** the strategies you used to solve Problem **1**.

5. **Explain** how you found numbers of boxes in order to solve Problems **2** and **3**.

Holt Mathematics

2-1 Rational Numbers

Comparing numbers with $\frac{1}{2}$ is useful in many situations.

Phil surveyed 317 voters and found
that 156 supported his reelection.
He reasoned the following way to
determine whether he had a majority.

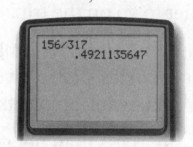

156/317
.4921135647

$156 \times 2 = 150 \times 2 + 6 \times 2$

$= 300 + 12$

$= 312$

Phil concluded that he did not have a majority.
Look at how Phil checked his estimate with a calculator.

**Double the numerator and compare it with the denominator
of each fraction to determine whether the fraction is greater
than or less than $\frac{1}{2}$. Check your work with a calculator.**

	Fraction	$> \frac{1}{2}$?	$< \frac{1}{2}$?
1.	$\frac{51}{101}$		
2.	$\frac{221}{425}$		
3.	$\frac{260}{513}$		
4.	$\frac{578}{1152}$		

Think and Discuss

5. Explain how you checked your work with a calculator.

Holt Mathematics

2-2 Comparing and Ordering Rational Numbers

You can use a number line to compare rational numbers. For example, the number line shows that $\frac{1}{8} < \frac{1}{4}$ because $\frac{1}{8}$ is to the left of $\frac{1}{4}$.

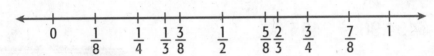

Use the number line to compare each pair of fractions by writing < or >. Then convert the fractions to decimals and use < or > to compare the decimals.

1. $\frac{5}{8} \ \square \ \frac{2}{3}$

2. $\frac{3}{8} \ \square \ \frac{1}{3}$

3. $\frac{7}{8} \ \square \ \frac{3}{4}$

4. $\frac{3}{8} \ \square \ \frac{2}{3}$

5. $\frac{3}{4} \ \square \ \frac{2}{3}$

6. $\frac{1}{4} \ \square \ \frac{1}{3}$

Think and Discuss

7. **Explain** how you know that $\frac{3}{8} < \frac{5}{8}$ without using the number line.

8. **Describe** how you could use the number line to compare $5\frac{3}{8}$ and $5\frac{1}{2}$.

Holt Mathematics

2-3 Adding and Subtracting Rational Numbers

Beckie has $454.96 in a checking account. She needs to pay bills in the amounts $25.95, $313.00, $45.76, and $87.95.

Beckie estimates the following:

Account: $454.96 ≈ $455.00

Bills: $25.95 ≈ $ 25.00
          $313.00 = $310.00
          $45.76 ≈ $ 45.00
          $87.95 ≈ $ 90.00
                    $470.00

Beckie determines that she does not have enough money in her account to pay her bills. She then checks her estimate with a calculator.

Estimate the solution to each expression in the table. Then use a calculator to solve.

		Estimate	Actual
1.	120 − 9.8		
2.	45 − 17.8 + 15.9 + 16.1 − 1.07		
3.	88.10 + 109.85		
4.	34.12 − 18.30 + 65.25		

Think and Discuss

5. **Describe** the estimation strategies you used.

Holt Mathematics

2-4 Multiplying Rational Numbers

To find the product of $\frac{1}{2} \cdot \frac{1}{3}$, use one color to color in $\frac{1}{2}$ of a square vertically. Use a second color to color in $\frac{1}{3}$ of the square horizontally. The product is represented by the area where the two colors intersect.

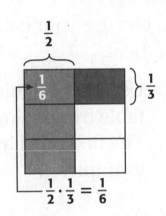

Use a similar model to find each product.

1. $\frac{1}{2} \cdot \frac{4}{6}$

2. $\frac{2}{3} \cdot \frac{3}{4}$

To multiply $2 \cdot 1\frac{1}{2}$, color in 2 squares across and $1\frac{1}{2}$ down.

$$2 \cdot 1\frac{1}{2} = 3$$

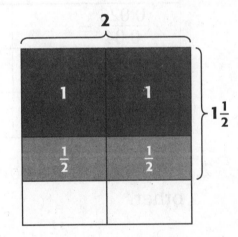

Use a similar model to find each product.

3. $3 \cdot 4\frac{1}{2}$

4. $8\frac{1}{4} \cdot 4$

Think and Discuss

5. **Explain** how the model shows that $2.5 \cdot 2.5 = 6.25$.

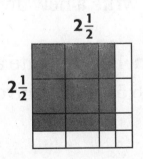

Holt Mathematics

2-5 Dividing Rational Numbers

You can use your calculator to investigate division of rational numbers.

1. Use your calculator to complete the table by dividing the rational numbers. The first one has been filled in as an example.

$\dfrac{16.8}{0.7} = 24$	$\dfrac{168}{7} =$	$\dfrac{1680}{70} =$
$\dfrac{4.1}{0.5} =$	$\dfrac{41}{5} =$	$\dfrac{410}{50} =$
$\dfrac{0.026}{0.04} =$	$\dfrac{0.26}{0.4} =$	$\dfrac{2.6}{4} =$
$\dfrac{75}{0.06} =$	$\dfrac{750}{0.6} =$	$\dfrac{7500}{6} =$
$\dfrac{16.32}{1.02} =$	$\dfrac{163.2}{10.2} =$	$\dfrac{1632}{102} =$

2. How are the division problems in each row related to each other?

Think and Discuss

3. **Show** how you can use the pattern in the first row of the table to write a new division problem whose quotient is equal to 24.

4. **Explain** how to write a division problem that is equivalent to $145 \div 16.5$ but that involves dividing by a whole number.

Holt Mathematics

2-6 Adding and Subtracting with Unlike Denominators

You can use models to show addition and subtraction of fractions with unlike denominators. Look at the models for $\frac{1}{2} + \frac{1}{3}$ and $\frac{1}{2} - \frac{1}{3}$.

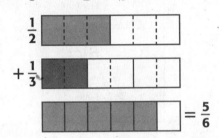

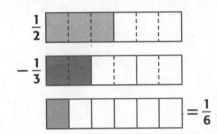

1. Draw a picture to show that $\frac{1}{2} + \frac{4}{5} = 1\frac{3}{10} = 1.3$.

2. Draw a picture to show that $\frac{1}{2} + \frac{2}{3} = \frac{7}{6} = 1\frac{1}{6}$.

3. Complete each addition problem. Simplify your answers.

   a. $\frac{7}{10} + \frac{4}{5}$ b. $\frac{3}{5} + \frac{3}{4}$

Think and Discuss

4. **Explain** how to add and subtract fractions with unlike denominators.

5. **Discuss** why the fraction that results from adding or subtracting fractions with unlike denominators has a different denominator than the fractions.

Holt Mathematics

2-7 Solving Equations with Rational Numbers

A box of cookies has 6 servings. Each serving contains 40.5 calories. How many calories are in the whole box?

40.5	40.5	$= 81$
40.5	40.5	$= 81$
40.5	40.5	$\dfrac{= 81}{243}$

Equation

$40.5 \cdot 6 = c$

$243 = c$

Estimate the solution for each equation. Then use a calculator to solve.

	Equation	Estimate	Actual
1.	$124.75 - x = 50$		
2.	$x + 16.9 = 15.5$		
3.	$0.6x = 15$		
4.	$\dfrac{x}{1.25} = 8$		

Think and Discuss_____

5. Describe how you estimated the solutions for the equations in Problems 1–4.

6. Discuss whether it is easier to estimate the solutions of some equations than the solutions of others.

Holt Mathematics

2-8 Solving Two-Step Equations

Felipe has a 750-gallon tank full of water on his ranch for his livestock. The livestock use 15 gallons of water per day, and no rain is expected to refill the tank during the dry season.

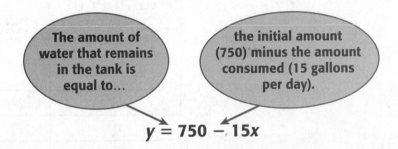

The amount of water that remains in the tank is equal to...

the initial amount (750) minus the amount consumed (15 gallons per day).

$$y = 750 - 15x$$

Use the equation $y = 750 - 15x$ to answer each question.

1. How much water is in the tank after the first day ($x = 1$)?

2. How much water is in the tank after 25 days ($x = 25$)?

3. What will be equal to zero in the equation $y = 750 - 15x$ when the tank is empty?

4. After how many days will the tank be empty?

Think and Discuss

5. **Discuss** what it means when 750 gallons (the initial amount) is exactly the same as $15x$ gallons (the amount consumed).

Holt Mathematics

3-1 Ordered Pairs

At the GasCo station, gasoline costs $3 per gallon. Let x represent the number of gallons of gasoline that you buy and let y represent the total cost of the gasoline.

1. Complete the table.

Number of Gallons, x	Total cost, y
1	$3
2	
3	
4	
5	
6	

2. Each row of the table contains an ***ordered pair.*** An ordered pair lists an x-value and its corresponding y-value. For example, the first row of the table contains the ordered pair (1, 3). Write the ordered pairs in the other rows of your table.

3. The equation $y = 3x$ gives the cost of x gallons of gasoline. Use the equation to find the cost of 12 gallons of gasoline.

4. Write the ordered pair from Problem 3.

Think and Discuss

5. Describe how you could find a new ordered pair for the above situation.

6. Explain whether you think the ordered pair (1, 3) is the same as (3, 1).

Holt Mathematics

3-2 Graphing on a Coordinate Plane

You are about to give directions to the locations labeled on the grid. The only restrictions are the following:

- You can move only horizontally (sideways) or vertically (up and down).

- Your first move should be horizontal. Use the directions *left* or *right*.

- Your second move should be vertical. Use the directions *up* or *down*.

Give directions to go to each location.

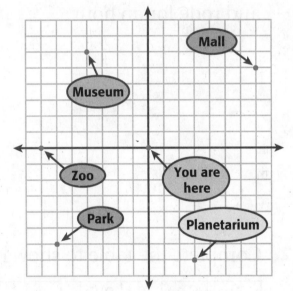

1. the museum

2. the park

3. the mall

4. the zoo

5. the planetarium

Think and Discuss

6. **Explain** how to write the directions for the locations in Problems 1–5 using *ordered-pair* notation. (*Hint:* the museum is at (−4, 6).)

Holt Mathematics

3-3 Interpreting Graphs and Tables

Graphs and tables can be used to represent many different situations.

Carmen rented a bicycle. The graph shows how far away she is from the rental site after each half hour of riding.

1. Use the graph to describe Carmen's trip. You can start the description like this: "Carmen left the bike shop and rode for an hour…"

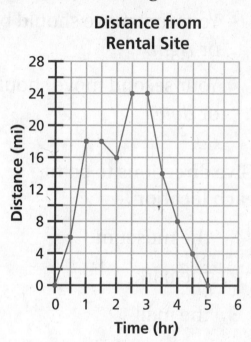

Distance from Rental Site

2. Complete the table to show her distance at each half hour.

Time (hr)	0	0.5	1	1.5	2	2.5	3	3.5	4	4.5	5
Distance											

Think and Discuss

3. **Explain** how you can tell when Carmen took a break.

4. **Determine** during which half hour Carmen covered the greatest distance.

Holt Mathematics

3-4 Functions

A *function* is a rule that assigns one *output* value for each *input* value. An input/output table is a convenient way of representing functions.

1. Complete the table by applying the rule to each input value.

Input x	Rule: $2000 - 25x$	Output y
0	$2000 - 25(0) = 2000$	2000
1		
2		
3		

2. Determine the rule that produces the following output values from the given input values.

Input x	Rule	Output y
2		0
4		2
6		4
8		6

Think and Discuss

3. Explain the relationship between the input values, a rule, and the output values.

4. Discuss whether the output values have to be different from the input values.

Holt Mathematics

3-5 Equations, Tables, and Graphs

Jerome is driving at a constant speed of 50 mi/h. You can use an equation, a table, and a graph to represent the distance d that Jerome travels in t hours.

1. Complete the table.

Time (hr), t	distance (mi), d
0.5	
1	50
1.5	
2	
2.5	
3	

2. Plot the points in your table and then connect the points to make a graph that shows Jerome's distance as a function of time.

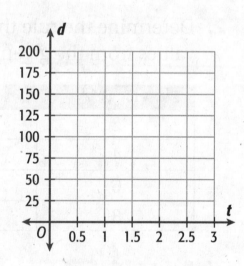

Think and Discuss

3. **Describe** the graph that you made.
4. **Explain** how distance is related to time. How can you use this to write an equation relating d and t?

Holt Mathematics

3-6 Arithmetic Sequences

Maya is a structural engineer. She is designing support structures for a bridge. As shown in the figures, the structures consist of regular beams (the horizontal and vertical beams) and cross beams (the diagonal beams).

| One section | Two sections | Three sections |

1. Maya makes a table showing the number of cross beams needed for various numbers of sections. Complete the table.

Number of Sections	1	2	3	4	5
Number of Cross Beams	2	4			

2. She also makes a table showing the number of regular beams needed for various numbers of sections. Complete the table.

Number of Sections	1	2	3	4	5
Number of Regular Beams	4	7			

Think and Discuss

3. **Describe** any patterns you notice in the table for the number of cross beams.

4. **Explain** how you could find the number of regular beams needed to make 6 sections based on the patterns in your table.

Holt Mathematics

4-1 Exponents

You can multiply $(-5) \cdot (-5) \cdot (-5) \cdot (-5) \cdot (-5) \cdot (-5)$
using exponents and a calculator.

The number -5 is a factor 6 times,
so you can write it as $(-5)^6$.

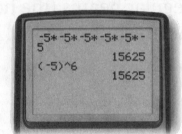

The expressions are equivalent because
they have the same value.

$(-5) \cdot (-5) \cdot (-5) \cdot (-5) \cdot (-5) \cdot (-5) =$
15,625 and $(-5)^6 = 15,625$

**Write each of the following using exponents. Then use a
calculator to find the value of the expression.**

1. $2 \cdot 2 \cdot 2 \cdot 2$

2. $8 \cdot 8 \cdot 8 \cdot 8 \cdot 8 \cdot 8$

3. $(-4) \cdot (-4) \cdot (-4)$

4. $(-5) \cdot (-5)$

5. $3 \cdot 3 \cdot 3 \cdot 3 \cdot 3 \cdot 3 \cdot 3 \cdot 3 \cdot 3$

6. $9 \cdot 9 \cdot 9$

Think and Discuss

7. Discuss whether 3^9 is the same as 9^3.

8. Explain why 4^6 is greater than 4^5.

Holt Mathematics

4-2 Look for a Pattern in Integer Exponents

Suppose the height of a magic plant doubles every hour, beginning with a height of 1 inch at the "zero hour."

Use the bar graph to think about how tall the plant was 1 hour before (−1) and 2 hours before (−2) the zero hour.

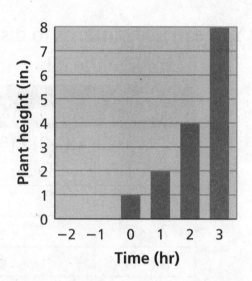

1. Draw and label a bar on the bar graph to show the height at hour −1 and hour −2. (*Hint:* The height doubles each hour.)

2. Use the graph and the pattern in the table to find the value of each exponential expression.

Time, x	Plant Height, 2^x
−3	$2^{-3} = \dfrac{1}{2^3} = \dfrac{1}{8}$
−2	
−1	
0	
1	
2	
3	

Think and Discuss

3. **Describe** the pattern in the bar graph.

4. **Describe** the pattern in the table.

Holt Mathematics

4-3 Properties of Exponents

You can use patterns to discover properties of exponents.

1. Complete the table.

Product of Powers	Write the Factors	Write As a Single Power
$3^2 \cdot 3^5$	$(3 \cdot 3) \cdot (3 \cdot 3 \cdot 3 \cdot 3 \cdot 3) =$ $3 \cdot 3 \cdot 3 \cdot 3 \cdot 3 \cdot 3 \cdot 3$	3^7
$4^3 \cdot 4^2$		
$7^4 \cdot 7^4$		
$5^2 \cdot 5^4$		

2. Look at the left-hand and right-hand columns of the table. What patterns do you notice?

3. Complete the table.

Product of Powers	Write the Factors	Write As a Single Power
$\dfrac{3^6}{3^2}$	$\dfrac{3 \cdot 3 \cdot 3 \cdot 3 \cdot 3 \cdot 3}{3 \cdot 3} = 3 \cdot 3 \cdot 3 \cdot 3$	3^4
$\dfrac{4^5}{4^3}$		
$\dfrac{5^4}{5^3}$		
$\dfrac{8^7}{8^3}$		

4. Look at the left-hand and right-hand columns of the table. What patterns do you notice?

Think and Discuss

5. **Explain** how you can use what you discovered to write $2^7 \cdot 2^{10}$ as a single power.

6. **Explain** how you can use what you discovered to write $\dfrac{2^{10}}{2^7}$ as a single power.

Holt Mathematics

4-4 Scientific Notation

1. Complete the table of values for the powers of ten.

Exponent	Power
–6	$10^{-6} =$
–5	$10^{-5} =$
–4	$10^{-4} =$
–3	$10^{-3} =$
–2	$10^{-2} = \dfrac{1}{10^2} = \dfrac{1}{10 \times 10} = 0.01$
–1	$10^{-1} = \dfrac{1}{10^1} = \dfrac{1}{10} = 0.1$
0	$10^0 = 1$
1	$10^1 = 10$
2	$10^2 = 10 \times 10 = 100$
3	$10^3 =$
4	$10^4 =$
5	$10^5 =$
6	$10^6 =$

Think and Discuss

2. **Discuss** the pattern you see in the table of values for powers of ten.

3. **Explain** how you know that $10^{-9} = 0.000000001$.

Holt Mathematics

4-5 Squares and Square Roots

The sequence shows the square numbers 1, 4, 9, and 16.

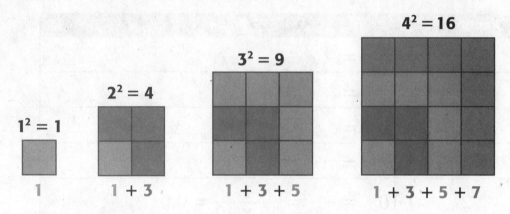

1. Draw a picture to show that $5^2 = 1 + 3 + 5 + 7 + 9$.

2. Add the odd numbers $1 + 3 + 5 + 7 + 9 + \cdots + 17 + 19$. What square number do you get?

3. The table starts with $11^2 = 1 + 3 + 5 + 7 + 9 + 11 + 13 + 15 + 17 + 19 + 21 = 121$. Complete the table by adding the next odd number to this sum.

11^2	12^2	13^2	14^2	15^2	16^2	17^2	18^2	19^2	20^2
121									

Think and Discuss

4. **Explain** how you can determine square numbers using sums of odd numbers.

5. **Demonstrate** that the value of 22^2 can be determined by adding odd numbers.

Holt Mathematics

4-6 Estimating Square Roots

Knowing the square numbers can help you estimate square roots.

1. Complete the table of squares.

1^2	2^2	3^2	4^2	5^2	6^2	7^2	8^2	9^2	10^2
1									

11^2	12^2	13^2	14^2	15^2	16^2	17^2	18^2	19^2	20^2
121									

Use the table of squares above to help you estimate each square root to the nearest tenth. Use a calculator to check your estimates. Round to two decimal places.

	Square Root	Estimate	Calculator
2.	$\sqrt{10}$		
3.	$\sqrt{20}$		
4.	$\sqrt{200}$		
5.	$\sqrt{300}$		
6.	$\sqrt{57}$		
7.	$\sqrt{130}$		

Think and Discuss

8. Discuss your strategy for estimating square roots.

Holt Mathematics

4-7 The Real Numbers

The Set of Real Numbers

Rational numbers can be written as fractions or as decimals that terminate or repeat.	*Irrational numbers* are decimals that do not terminate or repeat.
Examples: $\frac{1}{4}$ $\quad$ $5 = \frac{5}{1}$ $\sqrt{1.69} = 1.3$ $\frac{1}{6} = 1.\overline{6}$	Examples: $\sqrt{7} = 2.645751311\ldots$ $\pi = 3.141592654\ldots$

Classify each number as rational or irrational.

		Rational	Irrational
1.	$2\frac{1}{2}$		
2.	$\sqrt{24}$		
3.	$\sqrt{16}$		
4.	$7.\overline{7}$		
5.	$\sqrt{\frac{4}{9}}$		

Think and Discuss

6. **Explain** how to classify numbers as rational.

7. **Explain** how to classify numbers as irrational.

Holt Mathematics

4-8 The Pythagorean Theorem

The model shows a visual example of the Pythagorean Theorem.

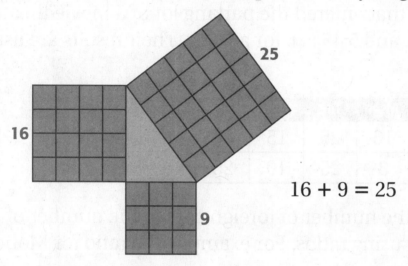

$16 + 9 = 25$

1. Find the lengths of the sides of the right triangle that is surrounded by the three squares.

2. What is the area of each square on each side of the triangle?

3. How does the length of each side of the triangle relate to the area of the corresponding square?

4. Can you form another example of the Pythagorean Theorem with the squares below?

Think and Discuss _____

5. **Explain** how you found the answer for Problem 4.

Holt Mathematics

5-1 Ratios and Proportions

Sam and Jill counted the number of foreign cars and the number of domestic cars that entered the parking lot of a movie theater between 5:30 P.M. and 5:45 P.M. for a week. Their results are listed in the table.

	Mon	Tue	Wed	Thu	Fri	Sat	Sun
Foreign	16	25	15	20	40	40	45
Domestic	8	25	10	25	20	50	45

They compared the number of foreign cars to the number of domestic cars by using ratios. For example, the ratio for Monday was $\frac{\text{foreign}}{\text{domestic}} = \frac{16}{8} = \frac{2}{1}$.

1. Write a ratio for each day.

2. Which ratios are greater than the Monday ratio?

3. Which ratios are less than the Monday ratio?

4. Which ratio is the same as the Monday ratio?

Think and Discuss

5. **Explain** what it means if two ratios are equal.

6. **Describe** how the ratios are different if you write them as $\frac{\text{domestic}}{\text{foreign}}$.

Holt Mathematics

5-2 Ratios, Rates, and Unit Rates

The bar graph shows the number of acres of wilderness burned each year from 1991 to 2000. Each bar represents a *unit rate*, because it shows the number of acres burned in *one* year.

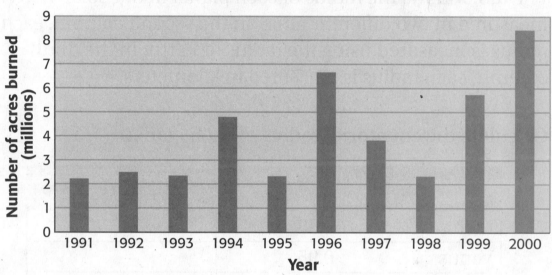

Acres of Wilderness Burned: 1991–2000

1. The National Interagency Fire Center reported that an average of 3,647,883 acres were burned per year from 1991 to 2000.

   a. Is the average also a unit rate?

   b. In what years were the number of acres burned above the average?

   c. In what years were the number of acres burned below the average?

Think and Discuss

2. **Explain** why a unit rate such as 40 miles per hour may be more useful than an equivalent rate such as 60 miles per 1.5 hours.

Holt Mathematics

5-3 Dimensional Analysis

The radius of a planet is called the *equatorial radius* and is the imaginary line between the center of the planet and a point on its equator.

The table shows the radius of each planet in our solar system measured in two different units. In the second column, each radius is measured using the radius of Earth. In the third column, each radius is measured in kilometers.

Calculate the equatorial radius of each planet.

	Planet	Equatorial Radius (number of Earth radii)	Equatorial Radius (km)
1.	Mercury	0.38	
2.	Venus	0.95	
3.	Earth	1	6378.14
4.	Mars	0.53	
5.	Jupiter	11	
6.	Saturn	9	
7.	Uranus	4	
8.	Neptune	4	
9.	Pluto*	0.19	

*designated as a dwarf planet in 2006

Think and Discuss

10. **Explain** how you found the equatorial radius of each planet.
11. **Discuss** whether using Earth's radius or the kilometer makes comparing the radii of the planets easier.

Holt Mathematics

5-4 Solving Proportions

The grid shows three rectangles aligned by a diagonal line.

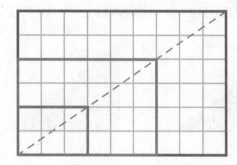

Complete the table. Use the grid to measure the length and width of each rectangle. Divide the length by the width and write this answer as a fraction in simplest form.

		Length	Width	$\dfrac{\text{Length}}{\text{Width}}$
1.	Blue rectangle			
2.	Red rectangle			
3.	Green rectangle			

Think and Discuss

4. **Compare** the fractions you found in Problems 1–3.

5. **Explain** how you wrote each fraction in simplest form.

Holt Mathematics

5-5 Similar Figures

Similar figures have the same shape, but not necessarily the same size. The two triangles shown below are similar.

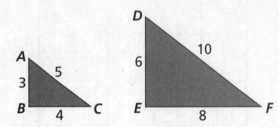

1. What appears to be true about each pair of angles below?

    ∠A and ∠D   ∠B and ∠E   ∠C and ∠F

2. Calculate each of the ratios shown below. What do you notice?

    length of *AB*   length of *BC*   length of *BC*
    ────────────   ────────────   ────────────
    length of *DE*   length of *EF*   length of *DF*

3. The two rectangles are similar to each other. What is true about the pairs of angles ∠J and ∠P, ∠K and ∠Q, ∠L and ∠R, and ∠M and ∠S?

    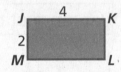

4. What is true about all the ratios
    length of *JK*   length of *KL*
    ────────────, ────────────,
    length of *PQ*   length of *QR*

    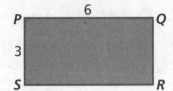

    length of *LM*   length of *MJ*
    ────────────, and ────────────?
    length of *RS*   length of *SP*

Think and Discuss

5. **Describe** what must be true when two polygons are similar to each other.

Holt Mathematics

5-6 Dilations

A *dilation* is an enlargement or a reduction of a figure.

For each figure, draw a new figure that has the same shape but is twice as large as the original.

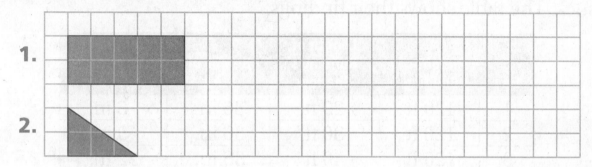

1.

2.

For each figure, draw a new figure that has the same shape but is half as large as the original.

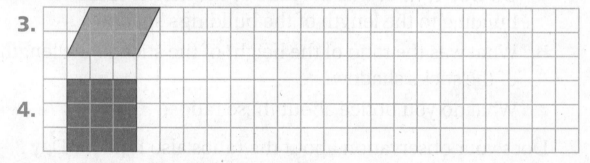

3.

4.

Think and Discuss

5. **Explain** how you maintained the same shape when enlarging the figures in Problems **1** and **2**.

6. **Explain** how you maintained the same shape when reducing the figures in Problems **3** and **4**.

Holt Mathematics

5-7 Indirect Measurement

Mr. Kelly's science class is tracking the motion of the sun during the year. To do so, they measure the shadow cast by a 120-foot building on four different days of the year. To check the measurements, the students also measure the shadow cast by a 36-inch stick. The table shows their findings.

Day	Height of Building	Length of Building's Shadow	Height of Stick	Length of Stick's Shadow
1	120 ft	20 ft	36 in.	6 in.
2	120 ft	30 ft	36 in.	9 in.
3	120 ft	40 ft	36 in.	12 in.
4	120 ft	80 ft	36 in.	24 in.

1. **a.** On Day 1, what was the ratio of the height of the building to the length of the building's shadow?

   b. What was the ratio of the height of the stick to the length of the stick's shadow?

   c. What do you notice about these ratios?

2. Does your observation about the ratios also hold for Day 2, Day 3, and Day 4?

Think and Discuss

3. **Describe** how you could find the length of the stick's shadow if you know that the length of the building's shadow is 60 ft.

4. **Explain** what must be true about the building's shadow on a day when the stick's shadow is 36 in.

Holt Mathematics

5-8 Scale Drawings and Scale Models

A coffee house rents the floor space modeled by the figure below.

☐ = 1 ft²

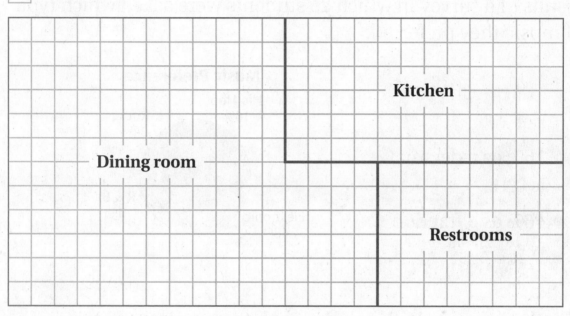

Count the number of squares to answer each question.

1. What is the actual area of the dining room?

2. What is the actual area of the kitchen?

3. What is the actual area of the restrooms?

Think and Discuss

4. **Show** a possible arrangement of square tables in the dining area if the tabletops each measure 2 ft by 2 ft.

5. **Discuss** real-world situations in which scale drawings are used.

Holt Mathematics

6-1 Relating Decimals, Fractions, and Percents

A *percent* is a ratio that compares a number to 100. Percents can be modeled on circle graphs. The circle graph below shows the results of a survey in which 25 students were asked which type of music they preferred.

- 16% means 16 per 100.

- 16% as a decimal is 0.16.

- 16% as a fraction is $\frac{16}{100} = \frac{4}{25}$.

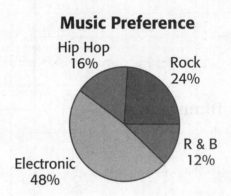

Music Preference

Hip Hop 16%
Rock 24%
R & B 12%
Electronic 48%

Use the percents in the circle graph to complete the table.

	Music Type	Percent	Decimal	Fraction
1.	Electronic	48%		
2.	Rock	24%		
3.	Hip Hop	16%		
4.	R & B	12%		

Think and Discuss

5. **Explain** how you wrote each percent as a decimal.

6. **Explain** how you wrote each decimal as a fraction.

Holt Mathematics

6-2 Estimate with Percents

You can use compatible numbers when estimating with percents. For example, to find 24% of 82, use 25% of 80.

$0.25 \cdot 80 = 20$
So 24% of 82 is approximately 20.

For each problem, estimate using compatible numbers. Then use a calculator to find the actual answer.

		Estimate	Actual
1.	Of 610 students, 34% prefer domestic cars over foreign cars.		
2.	In a survey, 19% of 152 people selected juice as their favorite drink.		
3.	A family wishes to leave a 15% tip on a $32.15 bill at a restaurant.		
4.	In a country with a population of 10,036,724, 48% are males.		

Think and Discuss

5. Discuss your strategies for estimating.

6. Explain how to avoid overestimating or underestimating.

Holt Mathematics

6-3 Finding Percents

The circle graph shows results from a mock election in which 40 students were polled.

Percent of Votes Received by Each Candidate

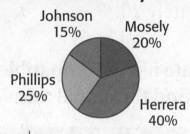

Johnson 15%
Mosely 20%
Phillips 25%
Herrera 40%

Complete the table to find the number of votes each candidate received.

	Candidate	Percent of Votes	Number of Votes
	Mosely	20%	$0.20 \cdot 40 = 8$
1.	Herrera	40%	
2.	Phillips	25%	
3.	Johnson	15%	

Think and Discuss

4. **Discuss** whether there is a clear winner in the election.
5. **Tell** which two candidates combined received the same total votes as another candidate. Did they also get the same percent of the votes as the other candidate?

Holt Mathematics

6-4 Finding a Number When the Percent is Known

A CD player is on sale for $15 off the original price. According to the advertisement, this is a 25% discount. You can use what you know about percents to figure out the original price of the CD player.

Think: 25% is $\frac{1}{4}$, so $15 must be $\frac{1}{4}$ of the original price.

You can model the situation as shown.

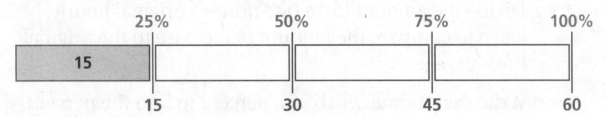

The original price is $60.

Draw a model to find the original price for each of the following.

1. Amount of discount: $8; percent of discount: 20%

2. Amount of discount: $12; percent of discount: $33\frac{1}{3}$%

3. Amount of discount: $17; percent of discount: 50%

Think and Discuss

4. **Explain** how you could find the value of a number given that 7 is 10% of the number.

5. **Describe** what must be true about a number if you know that 45 is 100% of the number.

Holt Mathematics

6-5 Percent Increase and Decrease

1. After his first year at a job, Andrew's original hourly wage of $7.95 increased to $9.54.

   a. Subtract the old hourly wage from the new hourly wage to find the amount by which Andrew's hourly wage increased.

   b. Divide the amount in **1a** by Andrew's original hourly wage to compare the amount of increase to the original hourly wage.

   c. Write the decimal in **1b** as a percent to find the percent increase in Andrew's hourly wage.

2. An electronics store offers a $100 discount on everything in the store that is priced between $500 and $900. Complete the table to determine the percent of decrease for each price.

Price	$500	$600	$700	$800	$900
Percent of Decrease	$\frac{100}{500} = \underline{\quad}\%$	$\frac{100}{600} =$	$\frac{100}{700} =$	$\frac{100}{800} =$	$\frac{100}{900} =$

Think and Discuss

3. **Explain** how you wrote the decimal in Problem **1c** as a percent.

4. **Compare** the percents of decrease in the table in Problem **2**.

Holt Mathematics

6-6 Applications of Percents

You often need to calculate percents when making a purchase.

Use a calculator to find the tax on each item and the total cost of the item including the tax.

	Item	Cost	Tax Rate	Tax	Total Cost = Cost + Tax
1.	CD	$13.95	8%		
2.	DVD	$24.99	8%		
3.	Headphones	$29.95	8%		

Use a calculator to find the total cost of each item.

	Item	Cost	Tax Rate	Total Cost = 1.08 · Cost
4.	CD	$13.95	8%	
5.	DVD	$24.99	8%	
6.	Headphones	$29.95	8%	

Think and Discuss

7. **Explain** how you calculated the tax on each item in Problems 1−3.

8. **Explain** why the total cost is the same whether you use the formula *total cost = cost + tax* or the formula *total cost = 1.08 · cost*.

Holt Mathematics

6-7 Simple Interest

Simple interest is the amount earned on money deposited in some savings accounts.

The interest your account earns is calculated using the formula $I = Prt$. For example, if you start a savings account with $100.00 (**principal P**) and your savings account pays 5% (**interest rate r**), the interest (I) you will have earned at the end of 1 year (**time t**) will be $0.05 \cdot 100 = \$5$. Your total balance at the end of 1 year will be $\$100.00 + \$5.00 = \$105.00$.

Complete the table.

	Savings	Interest Rate	Interest	Total Balance
	$100	5%	$0.05 \cdot 100 = \$5$	$100 + 5 = \$105$
1.	$200	6%		
2.	$300		$24	
3.	$500			$550

Think and Discuss

4. **Describe** your strategies for completing the table.
5. **Explain** how to use the formula $I = Prt$ when you need to find an interest rate.

7-1 Points, Lines, Planes, and Angles

In each group, one picture is different from the others. Identify the picture that is different and explain why it is different.

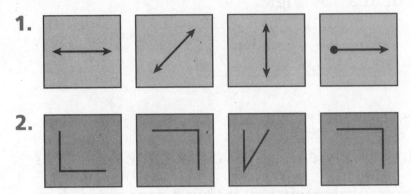

3. Look at the picture for each geometry term and write a real-world example for each term.

Term	Picture	Example
Point	•	
Segment	•——•	
Ray	•——→	
Line	←——→	
Angle		

Think and Discuss

4. **Explain** the difference between a *segment* and a *ray*.

5. **Explain** the difference between a ray and a *line*.

Holt Mathematics

7-2 Parallel and Perpendicular Lines

A carpenter cuts boards at different angles.

The board below is cut at a 45° angle.

1. Label the angles formed by the cut.

2. Which angles are *congruent,* or have the same measure?

3. Which angles are *supplementary,* or have measures that add to 180°?

The board below is cut at a 30° angle.

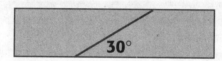

4. Label the angles formed by the cut.

5. Which angles are congruent?

6. Which angles are supplementary?

Think and Discuss

7. **Describe** the top and bottom edges of the boards above. What kinds of lines do they form?

8. **Describe** the angles formed after cutting the board at a 90° angle.

Holt Mathematics

7-3 Angles in Triangles

The figure below shows two parallel lines *m* and *n*, and a triangle.

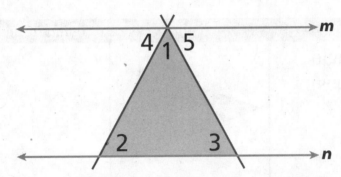

1. What is the sum of the measures of ∠4, ∠1, and ∠5?

2. Name an angle with the same measure as ∠4.

3. Name an angle with the same measure as ∠5.

4. What is the sum of the measures of ∠1, ∠2, and ∠3?

Think and Discuss

5. **Discuss** what would happen to the sum of the measures of ∠1, ∠2, and ∠3 if the triangle above were bigger.

6. **Explain** why each angle measure in an *equilateral triangle* is 60°. (In an equilateral triangle, the three sides are equal.)

Holt Mathematics

7-4 Classifying Polygons

Use a protractor to measure each angle in each figure.
Then find the sum of the angle measures.

	Polygon	Picture	Sum of Angles
1.	Triangle (3 sides)		
2.	Pentagon (5 sides)		
3.	Hexagon (6 sides)		

Think and Discuss

4. **Explain** how you know the sum of the angle measures in a square.

5. **Discuss** how you can use the two triangles to show that the sum of the angle measures in a four-sided figure is 360°.

Holt Mathematics

7-5 Coordinate Geometry

Quadrilaterals are figures with four sides. Use the coordinate grid to draw each of the following quadrilaterals. Label vertices with *A, B, C,* and *D.*

Quadrilateral	Graph
1. Square A (3, 2), B (3, 8), C (−3, 8), D (−3, 2)	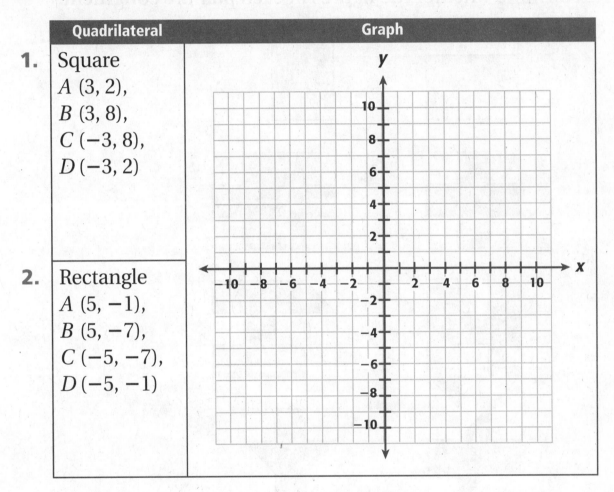
2. Rectangle A (5, −1), B (5, −7), C (−5, −7), D (−5, −1)	

Think and Discuss

3. Describe the relationship between the number of sides and the number of vertices of quadrilaterals.

4. Discuss whether the relationship you described in Problem **3** is also true for other polygons.

Holt Mathematics

7-6 Congruence

Congruent figures have the same size and shape.

Determine whether the figures in each pair are congruent.

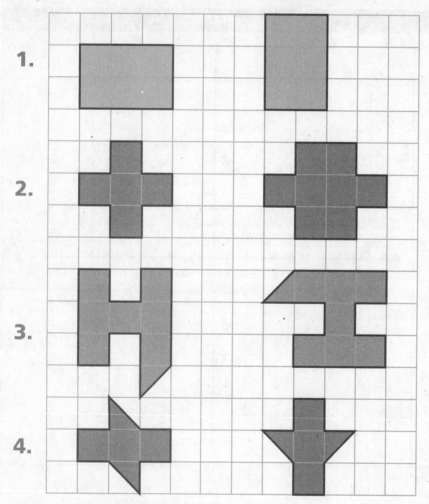

1.

2.

3.

4.

Think and Discuss

5. **Explain** how you decided whether the figures in Problems **1–4** were congruent or not.

6. **Find** two examples of objects that are congruent. What makes the two objects congruent?

Holt Mathematics

7-7 Transformations

1. Draw arrows from all the vertices (corners) of each original figure (blue) to the corresponding vertices of its *image* (red).

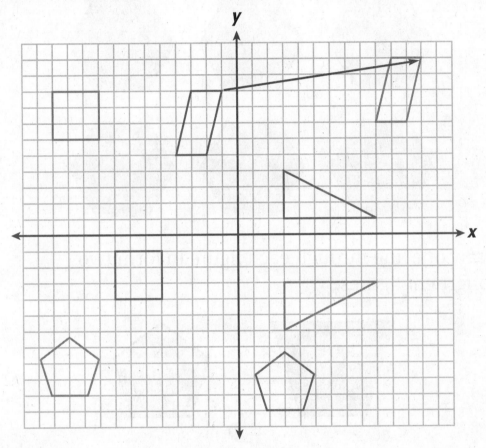

Think and Discuss

2. **Explain** whether the transformations in Problem **1** are congruent, which means that the image is the same shape and size as the original figure.

3. **Compare** the transformation of the triangle with the transformations of the other figures. How is it different? Find a name for this transformation.

Holt Mathematics

7-8 Symmetry

1. Draw one line through each hexagon to form two congruent figures. Use a different line for each hexagon.

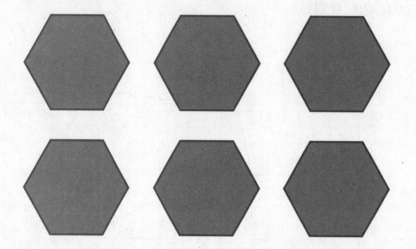

2. Draw one line through each figure to form two congruent figures.

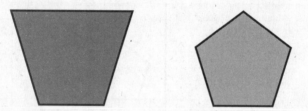

Think and Discuss

3. **Draw** a figure that cannot be cut into two congruent figures with one line.

4. **Describe** the characteristics of a figure that can be cut into two congruent figures.

Holt Mathematics

7-9 Tessellations

Tile floors come in many different designs. For each figure, determine whether or not it could be used as a floor tile. Draw a model to show a possible tiling pattern.

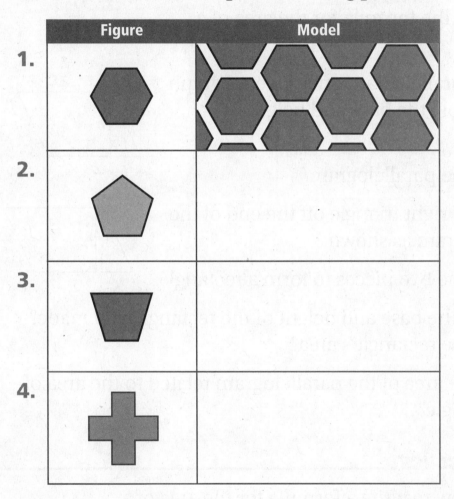

	Figure	Model
1.		
2.		
3.		
4.		

Think and Discuss

5. **Draw** a figure that cannot be used as a tile for a floor.

6. **Describe** the common characteristics that make some of the figures above suitable for floor tiles.

Holt Mathematics

8-1 Perimeter and Area of Rectangles and Parallelograms

Recall that the area of a rectangle is the product of its length times its width, or the product of its base times its height, *bh*. You can use this fact to develop the formula for the area of a parallelogram.

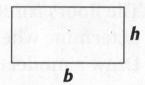

1. Draw a parallelogram on a sheet of graph paper. Label the base and height as shown.

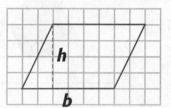

2. Cut out the parallelogram.

3. Now cut a right triangle off the end of the parallelogram as shown.

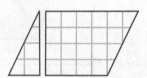

4. Arrange the two pieces to form a rectangle.

5. What are the base and height of the rectangle you made? What is the rectangle's area?

6. How is the area of the parallelogram related to the area of the rectangle?

Think and Discuss

7. **Explain** how to write a formula for the area of a parallelogram with base *b* and height *h*.

8. **Describe** how you can use your formula to find the area of this parallelogram.

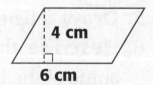

4 cm

6 cm

Holt Mathematics

8-2 Perimeter and Area of Triangles and Trapezoids

You can use what you know about the area of a parallelogram to develop the formula for the area of a triangle.

1. Fold a sheet of paper in half.

2. Draw a triangle on the folded paper. Label the base, b, and height, h, as shown.

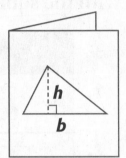

3. Cut out the triangle through both layers of paper. This will create a pair of congruent triangles.

4. Arrange the two triangles to form a parallelogram.

5. What are the base and height of the parallelogram you made? What is the parallelogram's area?

6. How is the area of one of the triangles related to the area of the parallelogram?

Think and Discuss

7. **Explain** how to write a formula for the area of a triangle with base b and height h.

8. **Describe** how you can use your formula to find the area of this triangle.

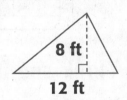

8 ft

12 ft

Holt Mathematics

8-3 Circles

First estimate the area of each circle by counting squares. Then square each circle's radius. Then compare each estimated area with the square of the radius by computing $\frac{A}{r^2}$.

		Radius	Estimated Area	r^2	$\frac{A}{r^2}$
1.		$r = 1$			
2.		$r = 2$			
3.		$r = 3$			
4.		$r = 4$			

Think and Discuss

5. **Discuss** how you can make a generalization about how to estimate the area of a circle if you know the radius.

Holt Mathematics

8-4 Drawing Three-Dimensional Figures

The figure shows an *isometric drawing* of a three-dimensional figure. Given this drawing, you can build the three-dimensional figure using two cubes.

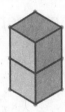

Use cubes to build each of the three-dimensional figures. Then tell how many cubes it takes to build each figure.

1.

2.

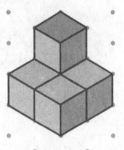

3.

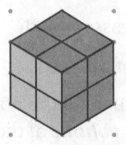

4.

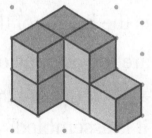

Think and Discuss

5. Discuss which of the three-dimensional figures in Problems 1–4 would look exactly the same when viewed from the front, top, and side.

6. Describe the top view of the figure in Problem 2.

Holt Mathematics

8-5 Volume of Prisms and Cylinders

The drawing below represents a box that has been unfolded and laid flat. The blue rectangles represent the bottom and the lid of the box. The green rectangles represent the sides of the box.

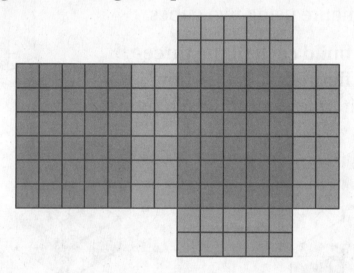

1. What are the dimensions of the bottom and the lid of the box?

2. What is the height of the box when it is assembled?

3. What are the dimensions of each side of the box?

4. How many 1-by-1-by-1 cubes will it take to fill the box when it is assembled? (*Hint:* This is the *volume* of the box.)

Think and Discuss _____

5. **Explain** how you can determine the volume of a box if you know the dimensions.

6. **Discuss** whether you could figure out the height of a box if you know the volume and the dimensions of the base.

Holt Mathematics

EXPLORATION

8-6 Volume of Pyramids and Cones

The tip of a sharpened pencil is shaped like a cone. How much of the pencil is lost after the tip is formed? To answer this question, you should know that the volume of a cone is $\frac{1}{3}$ the volume of the cylinder from which it was formed.

The tip of this pencil was formed out of a cylinder with a height of 0.8 cm and a diameter of 1cm. The cylinder had a volume of approximately 0.63 cm³.

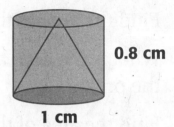

0.8 cm

1 cm

volume of cone $= \frac{1}{3}$ · volume of cylinder

volume of cone $= \frac{1}{3}$ · 0.63 = 0.21

Since the tip of the pencil has a volume of 0.21 cm³, 0.42 cm³ was lost when the tip of the pencil was formed.

Find the volume of each cone.

	Volume of Cylinder	Volume of Cone $= \frac{1}{3}$ · Volume of Cylinder
1.	66.9 in³	
2.	99 cm³	
3.	108 in³	

Think and Discuss

4. Name familiar objects that are shaped like a cone.

5. Explain how to estimate the volume of one of the cone-shaped objects you named in Problem 4.

Holt Mathematics

8-7 Surface Area of Prisms and Cylinders

The *surface area* of a three-dimensional figure is the sum of the areas of all the surfaces of the figure. Follow these steps to find the surface area of the prism.

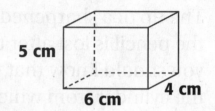

5 cm
4 cm
6 cm

1. Find the area of the bottom face of the prism.

2. What is the total area of the top and bottom faces of the prism?

3. Find the area of the front face of the prism.

4. What is the total area of the front and back faces of the prism?

5. Find the area of one of the side faces of the prism.

6. What is the total area of both side faces of the prism?

7. Add the areas in Steps 2, 4, and 6 to find the surface area.

Think and Discuss

8. **Explain** how you can find the surface area of a rectangular prism.

9. **Describe** a shortcut you can use to find the surface area of a cube.

Holt Mathematics

8-8 Surface Area of Pyramids and Cones

Look at how the figures are built to answer each question.

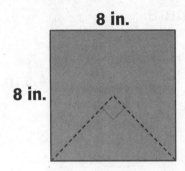

8 in.

8 in.

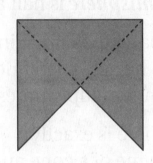

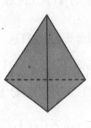

1. What is the surface area of the pyramid without a base?

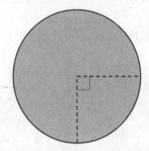

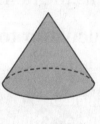

Area of circle $= 16\pi$ in^2.

2. What is the surface area of the cone without a base?

Think and Discuss

3. Explain what you would need to do to find the surface area of the pyramid and cone including the base.

Holt Mathematics

EXPLORATION

8-9 Spheres

A basketball is an example of a sphere. The radius of a basketball is about 4.5 inches. A *hemisphere* is half a sphere.

To find the volume of a basketball, imagine the ball is sliced into two halves. Then find the volume of one half and multiply times 2.

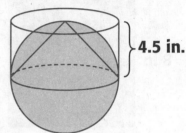

4.5 in.

The volume of a hemisphere is exactly halfway between the volume of a cone and the volume of a cylinder that both have the same radius r as the hemisphere and a height equal to r.

Use a calculator to find the volume of the following spheres.

	Volume of Cylinder Sitting on Top Half of Sphere	Volume of Cone = $\frac{1}{3}$ · Volume of Cylinder	Volume of Hemisphere = Halfway Between Vol. of Cylinder and Vol. of Cone	Volume of Sphere = Volume of Hemisphere · 2
1.	63.62 in^3	$\frac{1}{3}$ · 63.62 = 21.21 in^3	$\frac{63.62 + 21.21}{2}$ = 42.42 in^3	42.42 · 2 = 84.84 in^3
2.	70 in^3			
3.	27 in^3			
4.	123 in^3			

Think and Discuss

5. **Describe** how you could find the volume of the solid at right. The height of the dome is the same as the height of the cylinder, which is equal to the radius of the base.

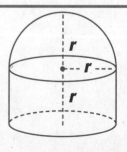

r

r

r

Holt Mathematics

8-10 Scaling Three-Dimensional Figures

A rectangular box has a volume of 2 cubic units. If one of its dimensions is doubled, the volume of the box is doubled. If all three dimensions are doubled, the volume is 8 times the volume of the original box.

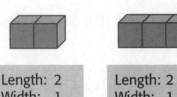

Length: 2
Width: 1
Height: 1
Volume:
$2 \cdot 1 \cdot 1 = 2$

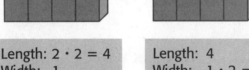

Length: $2 \cdot 2 = 4$
Width: 1
Height: 1
Volume:
$4 \cdot 1 \cdot 1 = 4$

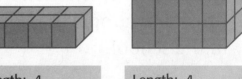

Length: 4
Width: $1 \cdot 2 = 2$
Height: 1
Volume:
$4 \cdot 2 \cdot 1 = 8$

Length: 4
Width: 2
Height: $1 \cdot 2 = 2$
Volume:
$4 \cdot 2 \cdot 2 = 16$

A box has a length of 3 in., a width of 2 in., and a height of 1 in. Multiply the dimensions of the box by each scale factor. Then find the volume.

	Scale Factor	Length (ℓ)	Width (w)	Height (h)	Volume $\ell \cdot w \cdot h$
	2	$3 \cdot 2 = 6$ in.	$2 \cdot 2 = 4$ in.	$1 \cdot 2 = 2$ in.	$6 \cdot 4 \cdot 2 = 48$ in^3
1.	3				
2.	4				
3.	5				

Think and Discuss

4. **Discuss** how each scale factor affects the volume of the rectangular box.

5. **Predict** what would happen to the volume if you were to multiply each dimension of the box by a scale factor of 0.5.

Holt Mathematics

9-1 Samples and Surveys

A survey is being conducted to determine what items should be offered at the school snack bar. Instead of including every student in the survey, a method of selecting a fair representation of all students must be used.

1. Determine which method of selecting students might provide a sample that fairly represents all of the students in the school.

   Method 1: selecting students whose last names begin with *A* or *B*

   Method 2: selecting students who purchased snack bar items in the past week

   Method 3: selecting all eighth-grade students

   Method 4: selecting one-fourth of all students at random

   Method 5: selecting the first 100 students to come to the office to take the survey

2. Explain one way to make the method you chose in Problem 1 even more likely to give a representative sample.

Think and Discuss

3. **Explain** your reasons for selecting the method you chose in Problem 1.

Holt Mathematics

9-2 Organizing Data

1. Students in a class completed a survey in which they were asked how many hours they watched TV during one month. Their responses are shown below.

15	33	10	25	8
72	21	2	30	21
10	20	36	6	16
27	14	47	25	14
33	14	30	15	25

a. Write the numbers in order from least to greatest.

b. Describe how the numbers are spread out.

c. One student from the class was absent on the day of the survey. If the student were asked the same question, what number of hours would you guess might be close to the student's answer?

Think and Discuss

2. **Explain** how writing the numbers in order from least to greatest helps to organize data.

3. **Describe** another way to organize the data using a table.

Holt Mathematics

9-3 Measures of Central Tendency

1. Students in a class completed a survey in which they were asked what their heights in inches are. Their responses are shown below.

49	53	60	55	48
72	65	66	58	68
75	65	64	57	59
61	67	64	58	62
63	59	61	55	65

   a. Write the numbers in order from least to greatest.

   b. What number appears most often?

   c. What number is in the middle of the data set?

   d. Add the numbers and divide the total by 25 to find the mean.

Think and Discuss

2. **Discuss** which number from Problems **1b, 1c,** and **1d** best represents the entire set of numbers. Why?

3. **Describe** a set of numbers arranged from least to greatest in which the middle number is not close to the value of the mean.

Holt Mathematics

9-4 Variability

When placed on a number line, values in a data set can be spread out or clustered together.

1. Order the data sets from the ones you think are the least spread out to the most spread out.

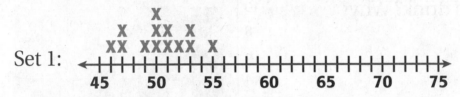

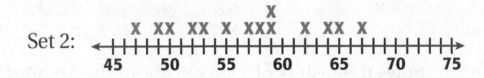

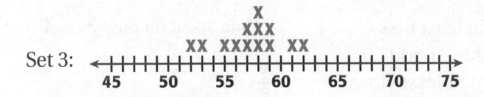

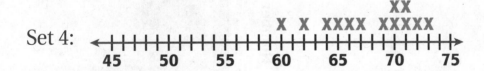

Think and Discuss

2. **Discuss** the strategies you used to order the sets according to the spread of the data.

3. **Explain** how the spread of set 1 would change if the value 72 were added.

Holt Mathematics

9-5 Displaying Data

The bar graph shows the results of a survey about students' lunch beverage preferences.

1. Do at least half the students prefer sports drink? Why?

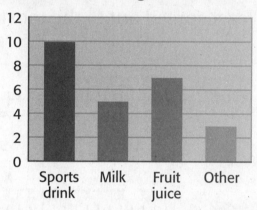

Students' Beverage Preferences

The histogram shows the results of a survey about the amount of time in hours per week that adults spend on the Internet.

2. Did at least half say they spend 20 hours or less? How do you know?

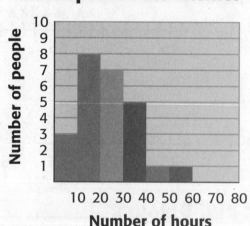

Time Spent on the Internet

Think and Discuss

3. **Compare** the types of data displayed by the bar graph with the types of data displayed by the histogram.

Holt Mathematics

9-6 Misleading Graphs and Statistics

The bar graph shows the percent of flavored water drinkers in a survey who prefer each brand of flavored water.

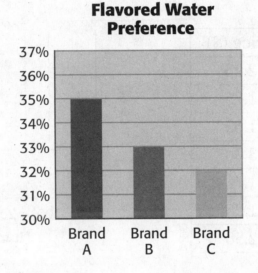

Flavored Water Preference

1. According to the survey, which brand is preferred the most?

2. How would the graph look if the percent scale started at 0% instead of 30%?

3. Why might someone want to display a graph like this?

Think and Discuss

4. **Explain** how the graph's appearance can be misleading.
5. **Discuss** situations that might be likely to involve misleading graphs or statistics.

Holt Mathematics

9-7 Scatter Plots

1. The table shows the numbers of pages in some paperback books and the books' prices. Plot the data on the graph provided.

Pages	Price ($)
300	2.25
200	1.75
130	1.65
450	3.00
180	1.75
75	1.25
250	2.50

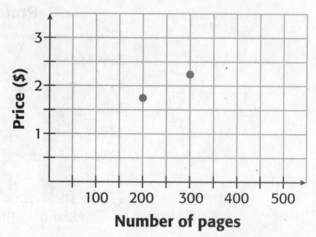

2. The table shows the number of hours some people spent outside and spent watching TV in one day. Plot the data on the graph provided.

TV hours	Outside hours
2.0	1.5
0.5	3.5
1.5	1.5
1.0	2.5
2.5	0.5
1.5	2.0
2.0	0.5

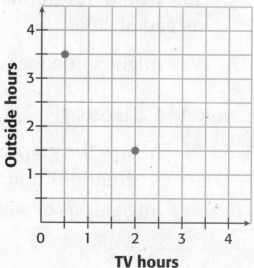

Think and Discuss

3. **Explain** what the shape of a graph tells you about the data being displayed.

Holt Mathematics

9-8 Choosing the Best Representation of Data

Some students sell water bottles at football games to raise money for their club. They want to know how many cases of water bottles to order, so they look at graphs showing the sales for last year.

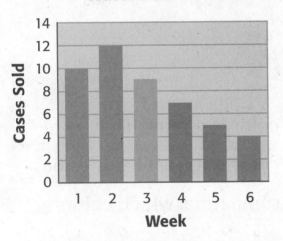

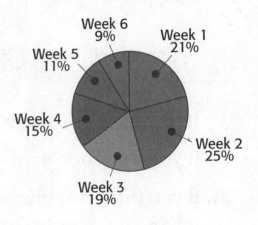

1. According to the bar graph, how many cases of water bottles were sold over the six-week period?

2. According to the circle graph, what percentage of the water bottles were sold in the first two weeks?

Think and Discuss

3. **Describe** the advantages of displaying this data in a bar graph.

4. **Describe** the advantages of displaying this data in a circle graph.

Holt Mathematics

10-1 Probability

A plastic container contains 10 marbles: 4 red, 3 yellow, 2 green, and 1 blue. An experiment consists of shaking the container and then drawing a number of marbles.

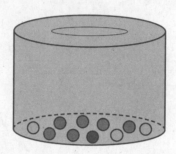

1. If you draw one marble, which color is it most likely to be?

2. If you draw one marble, which color is it least likely to be?

3. If you draw two marbles at the same time, which colors might you draw two of?

4. If you repeatedly draw one marble and return it to the container before the next draw, which color do you think you would draw most often?

Think and Discuss

5. **Discuss** real-world situations in which probability is used.

6. **Explain,** using the marble experiment, why the probability of drawing a blue marble is $\frac{1}{10}$. (*Hint:* Compare the number of blue marbles to the total number of marbles.)

Holt Mathematics

10-2 Experimental Probability

Use a plastic lid to make a spinner such as the one below. Color $\frac{1}{2}$ of the lid red, $\frac{1}{4}$ yellow, and the other $\frac{1}{4}$ green. Use a paper clip to secure the spinner to the center.

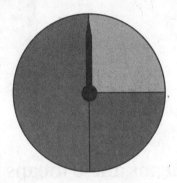

1. Spin the spinner 20 times, and record the results in the table.

Color	Number of Spins
Red	
Yellow	
Green	

   a. Write a ratio to compare the number of spins for each color to 20, the total number of spins.

   b. Add the three ratios in **a.** What is this sum equal to?

Think and Discuss

2. **Discuss** which color had the largest ratio in Problem **1a.**

3. **Explain** whether the sum in Problem **1b** is what you expected.

Holt Mathematics

10-3 Use a Simulation

You can use a calculator to simulate flipping a coin 30 times.

- Press the **MATH** key.

- Select **PRB.**

- Select **5:randInt(.**

- Key in 0, 1, 6.

- Press the **ENTER** key five times.

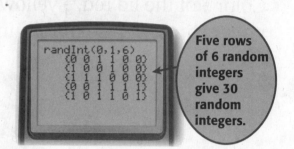

Five rows of 6 random integers give 30 random integers.

Assume that 1 means heads and 0 means tails to answer each question.

1. How many heads did you get in the 30 simulated flips?

2. How many tails did you get in the 30 simulated flips?

3. What is the experimental probability of flipping heads? (*Hint:* Write a fraction that compares the number of heads to 30 flips.)

4. What is the experimental probability of flipping tails? (*Hint:* Write a fraction that compares the number of tails to 30 flips.)

5. Use a calculator to simulate 100 flips, and compute the experimental probability of flipping heads.

Think and Discuss

6. **Explain** how to perform simulations on a calculator.

7. **Discuss** what you think would happen to the experimental probability as you increase the number of simulated flips.

Holt Mathematics

10-4 Theoretical Probability

The *theoretical probability* of an event tells you the probability of the event without your having to conduct an experiment.

For example, the experiment of rolling two dice and adding the two numbers that each die shows does not have to be conducted to know the possible sums of numbers.

	2	3	4	5	6	7
	3	4	5	6	7	8
	4	5	6	7	8	9
	5	6	7	8	9	10
	6	7	8	9	10	11
	7	8	9	10	11	12

1. Use the number of times each sum occurs to complete the table.

Sum	2	3	4	5	6	7	8	9	10	11	12
Outcomes	1	2									
Theoretical Probability	$\frac{1}{36}$	$\frac{2}{36}$									

Think and Discuss

2. **Explain** which sum is most likely to occur.

Holt Mathematics

10-5 Independent and Dependent Events

When the occurrence of one event affects the probability of a second event, the events are *dependent.* Otherwise, they are *independent.*

Classify each pair of events as dependent or independent.

Events		Independent	Dependent
1.	**Event A:** Drawing a 3 from a deck without replacing it **Event B:** Drawing another 3 from the same deck.		
2.	**Event A:** Tossing heads on a flip of a coin **Event B:** Tossing heads on a second flip of the same coin		
3.	**Event A:** Running between 5:00 P.M. and 5:30 P.M. **Event B:** Drinking water between 5:30 P.M. and 6:00 P.M.		
4.	**Event A:** Drawing a blue marble and putting it back in a bag that contains 6 blue marbles and 4 red marbles **Event B:** Drawing another blue marble from the same bag		

Think and Discuss

5. Discuss real-world examples of dependent events and independent events.

Holt Mathematics

10-6 Making Decisions and Predictions

Kate sells shirts and caps that have a school logo on them. Both the shirts and the caps are available in red or black. The table shows the number of shirts and caps that Kate sold during her first week of business.

	Red	Black
Shirts	125	75
Caps	60	40

1. What was the total number of shirts sold during the first week?

2. What fraction of the shirts that were sold were red shirts?

3. Kate assumes that future sales will be similar to the first week's sales. She plans to order 2000 more shirts. How many of these should be red shirts?

4. In general, when Kate sells a shirt, what is the probability that it is a red shirt?

5. What was the total number of caps sold during the first week?

6. What fraction of the caps that were sold were black caps?

7. Kate places an order for 5000 more caps. How many of these should be black caps?

Think and Discuss

8. **Explain** how you determined the number of red shirts that Kate should order in Problem 3.

9. **Explain** how you determined the number of black caps that Kate should order in Problem 6.

Holt Mathematics

10-7 Odds

When you find the *odds in favor* of an event, you compare the number of favorable outcomes to the number of unfavorable outcomes. With *probability,* you compare the number of favorable outcomes to the total number of outcomes.

The plastic container at right contains 10 marbles: 4 red, 3 yellow, 2 green, and 1 blue. The odds in favor of drawing a green marble are 2 to 8, because 2 marbles are green and 8 marbles are not green.

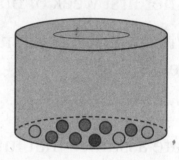

Use the definitions of *odds in favor* and *probability* to complete the table.

		Probability	Odds in Favor
	Drawing a green marble	$\frac{2}{10} = \frac{1}{5}$	2:8
1.	Drawing a red marble		
2.	Drawing a yellow marble		
3.	Drawing a blue marble		

Think and Discuss

4. **Explain** the difference between odds and probability.

Holt Mathematics

10-8 Counting Principles

A movie theater sells popcorn in small, medium, or large containers. Each size is also available for regular or lightly buttered popcorn.

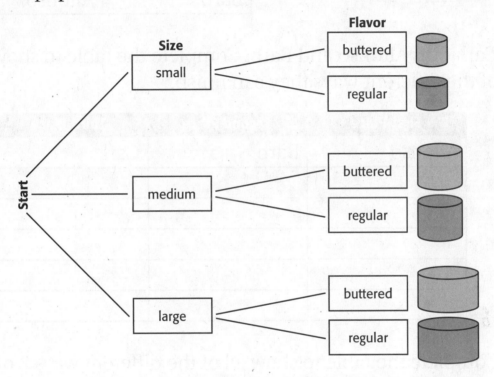

1. How many possible options are there for buying popcorn at the movie theater?

2. How many options are there if the theater adds two new flavors—cheddar cheese and caramel?

Think and Discuss

3. **Explain** how you counted the number of options in Problem 2.

4. **Discuss** whether the number of options in Problem 1 would be different if you were to start with flavor followed by size.

Holt Mathematics

10-9 Permutations and Combinations

Alfonso and Barb race against each other in a 5K race. There are two ways they can finish.

First Place	Second Place
Alfonso	Barb
Barb	Alfonso

1. Carl joins Alfonso and Barb. Complete the table to show all of the different ways they can finish.

	First Place	Second Place	Third Place
	Alfonso	Barb	Carl
a.			
b.			
c.			
d.			
e.			

2. Complete the table to show all of the different ways 2 of the 3 runners can finish in a tie for first place.

	First Place (tie)	Second Place
	Alfonso and Barb	Carl
a.		
b.		

Think and Discuss

3. **Explain** how you could find the number of ways 5 runners could finish.

Holt Mathematics

11-1 Simplifying Algebraic Expressions

Like terms are combined separately from unlike terms.
Look at how the like terms are combined in the diagram.

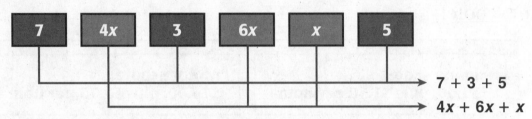

| 7 | 4x | 3 | 6x | x | 5 |

$7 + 3 + 5$

$4x + 6x + x$

Express the perimeter of each figure with all the like terms combined.

1.

4

5   5

x

perimeter = _____

2.

x   10

x

perimeter = _____

3.

7

2x   2x

7

perimeter = _____

4.

8   8

3x

perimeter = _____

Think and Discuss

5. Explain what it means to combine like terms.

6. Describe a real-world situation in which you combine like terms.

Holt Mathematics

11-2 Solving Multi-Step Equations

José and Rebecca want to rent sports equipment at the city park. José wants to rent inline skates, and Rebecca wants to rent a motor scooter.

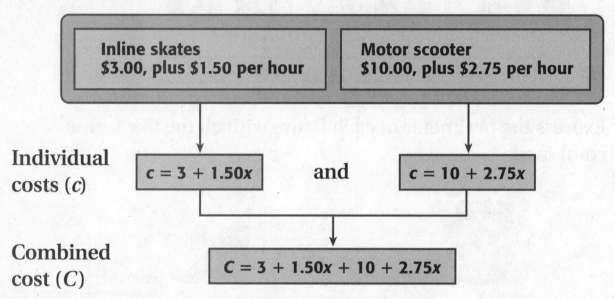

Inline skates
$3.00, plus $1.50 per hour

Motor scooter
$10.00, plus $2.75 per hour

Individual costs (c)

$c = 3 + 1.50x$ and $c = 10 + 2.75x$

Combined cost (C)

$C = 3 + 1.50x + 10 + 2.75x$

Use the equation $C = 3 + 1.50x + 10 + 2.75x$ to answer each question.

1. How much would José and Rebecca pay for 2 hours' rental?

2. How much would José and Rebecca pay for 3 hours' rental?

3. If their total rental cost is $38.50, for how many hours did they use the equipment?

Think and Discuss

4. **Explain** how to combine like terms in the equation $C = 3 + 1.50x + 10 + 2.75x$.

5. **Discuss** whether it makes sense to combine like terms when each item of sports equipment is rented for a different number of hours.

Holt Mathematics

11-3 Solving Equations with Variables on Both Sides

Samara wants to rent a kayak. Two rental places offer the following deals.

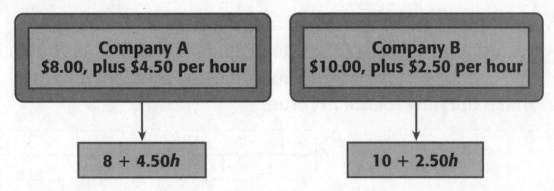

Company A	Company B
$8.00, plus $4.50 per hour	$10.00, plus $2.50 per hour

| $8 + 4.50h$ | $10 + 2.50h$ |

Use the expressions above to solve each problem.

1. Find the cost of a 1-hour rental from company A ($h = 1$).

2. Find the cost of a 1-hour rental from company B ($h = 1$).

3. Find the cost of a 2-hour rental from company A ($h = 2$).

4. Find the cost of a 2-hour rental from company B ($h = 2$).

5. Find the number of hours that a kayak would need to be rented from both companies to make the costs equal by solving $8 + 4.50h = 10 + 2.50h$ for h. (*Hint:* Combine like terms first.)

Think and Discuss

6. **Explain** how you solved the equation in Problem 5.

7. **Discuss** what it means to set $8 + 4.50h$ equal to $10 + 2.50h$.

Holt Mathematics

11-4 Solving Inequalities by Multiplying or Dividing

You can discover an important property of inequalities by looking for patterns.

1. Multiply both sides of each inequality by 2. Write the resulting numbers under the original inequality as shown in **1a.** Then insert the correct inequality symbol, < or >, to make the new inequality true.

a.

3	>	2
6		4

b.

−4	<	−1

c.

2	>	−3

2. Multiply both sides of each inequality by −2. Write the resulting numbers under the original inequality as shown in **2a.** Then insert the correct inequality symbol, < or >, to make the new inequality true.

a.

3	>	2
−6		−4

b.

−4	<	−1

c.

2	>	−3

Think and Discuss

6. **Describe** any patterns you notice.

7. **Explain** what happens when you multiply both sides of an inequality by a negative number.

Holt Mathematics

11-5 Solving Two-Step Inequalities

Rosa is offered two telephone service options when she buys her new cell phone.

Cell phone option A	Cell phone option B
$25, plus 12.5¢ per minute	$90, plus unlimited minutes

1. Complete the table to compare the costs under each option for the given number of minutes.

Minutes	Cost Under Option A	Cost Under Option B
220	$25 + 0.125 \cdot 220 = \$52.50$	$90
320		
420		
520		
620		

2. Solve the inequality $25 + 0.125x < 90$. What does the solution tell you about option A and option B?

3. Solve the inequality $25 + 0.125x > 90$. What does the solution tell you about option A and option B?

Think and Discuss

4. **Explain** how you solved the inequalities in Problems 2 and 3.

Holt Mathematics

11-6 Systems of Equations

You can compare Celsius temperatures with Fahrenheit temperatures by graphing each formula as an equation using the variables x and y.

To convert °C to °F

$F = \frac{9}{5} \cdot C + 32$

$\downarrow \qquad \downarrow$

$y = \frac{9}{5} \cdot x + 32$

To convert °F to °C

$C = \frac{5}{9} \cdot (F - 32)$

$\downarrow \qquad \downarrow$

$y = \frac{5}{9} \cdot (x - 32)$

The calculator screens below show how to graph the two equations.

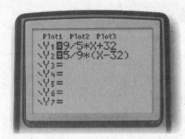

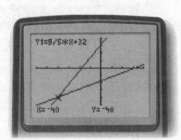

1. Which temperature is the same in both degrees Celsius and degrees Fahrenheit?

2. Substitute the temperature you found in Problem **1** for x in each equation above.

Think and Discuss

3. **Discuss** what happened when you substituted the temperature found in Problem **1** in each equation.

4. **Explain** what the point of intersection of the two graphs represents.

Holt Mathematics

12-1 Graphing Linear Equations

The calculator screens show the graphs of $y = x + 1$ and $y = x + 3$.

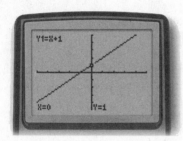 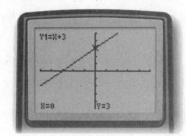

1. Why is the graph of $y = x + 3$ higher along the y-axis than the graph of $y = x + 1$?

The calculator screens show the graphs of $y = x + 2$ and $y = 2x + 2$.

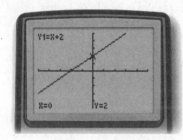 

2. Why is the graph of $y = x + 2$ flatter than the graph of $y = 2x + 2$?

Think and Discuss

3. Discuss how the two graphs for Problem 1 are similar and different.

4. Discuss how the two graphs for Problem 2 are similar and different.

Holt Mathematics

12-2 Slope of a Line

Joseph plots a graph of his 300-mile trip to the coast.

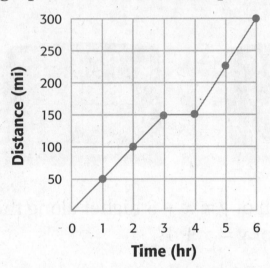

1. Calculate his average speed for the first 3 hours.
   (*Hint:* Divide the distance traveled during the first
   three hours by 3.)

2. What happens between hours 3 and 4?

3. Calculate his average speed for the last 2 hours.
   (*Hint:* Divide the distance traveled during the last
   two hours by 2.)

Think and Discuss

4. **Discuss** how you could find Joseph's average speed between
   hours 3 and 4.

5. **Discuss** how average speed is related to the shape of the graph.

Holt Mathematics

12-3 Using Slopes and Intercepts

On the graph below, 40 is called the *y-intercept* and 8 is called the *x-intercept.*

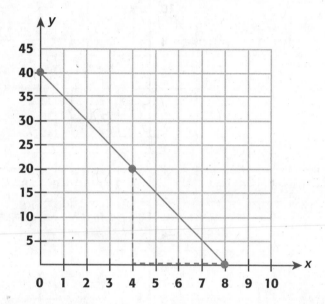

1. How do the rise and run compare to the intercepts?

2. Find the slope of the line by dividing the *y*-intercept by the *x*-intercept and then changing the sign of the quotient. Express the slope as a fraction in simplest form.

3. Consider the triangle drawn along the dashed lines below the line. Divide the vertical distance (20) by the horizontal distance (4), change the sign, and then compare this fraction with the fraction you found in Problem **2.**

Think and Discuss

4. **Explain** how to calculate the slope of a line using the *x*- and *y*-intercepts.

5. **Explain** how to calculate the slope of a line using any two points on the line.

Holt Mathematics

12-4 Point-Slope Form

The graph shows the point (2, 4).

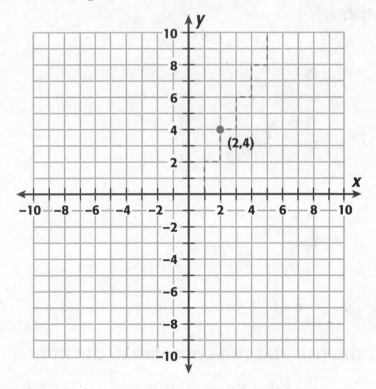

1. Draw a line that rests on the "stairs" drawn with a dashed line and that passes through the point (2, 4).

2. Calculate the slope of the line by dividing the vertical distance of each step by the horizontal distance of each step of the stairs.

3. Use the **point-slope form** $y - y_1 = m(x - x_1)$ to find the equation for the line. (*Hint:* $y_1 = 4$, $x_1 = 2$, and $m =$ the slope you found in Problem 2.)

Think and Discuss

4. **Explain** why the equation $y - y_1 = m(x - x_1)$ is called the point-slope form of a line.

Holt Mathematics

12-5 Direct Variation

In a *direct variation,* when one quantity increases or decreases, the other quantity does the same. The table below shows the number of stamps *x* and the price *y* for each number.

Notice that when the number of stamps is doubled, the price is also doubled.

Stamps x	1	2	3	4	5	6
Price y	$0.37	$0.74	$1.11	$1.48	$1.85	$2.22

1. Use each pair of values (*x, y*) in the table to complete the graph.

2. To find the *constant of proportionality,* divide each price in the table by each number of stamps.

3. What feature of the graph tells you that the graph is of a direct variation?

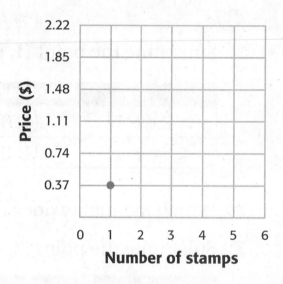

Think and Discuss

4. **Explain** the relationship between a constant of proportionality and the slope of the line.

5. **Give** a real-world example of a direct variation.

Holt Mathematics

12-6 Graphing Inequalities in Two Variables

The linear equation $y = 2x + 1$ is graphed at right.

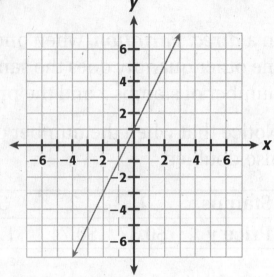

1. Substitute the point (1, 6) into each inequality.

	Point Above Line	Substituted into Inequality
$y > 2x + 1$	(1, 6)	
$y < 2x + 1$	(1, 6)	

2. Which inequality does the point (1, 6) satisfy?

3. Substitute the point (2, −4) into each inequality.

	Point Below Line	Substituted into Inequality
$y > 2x + 1$	(2, −4)	
$y < 2x + 1$	(2, −4)	

4. Which inequality does the point (2, −4) satisfy?

Think and Discuss

5. **Discuss** which points (those above or those below the line) you would expect to satisfy the inequalities $y > 2x + 1$ and $y < 2x + 1$.

Holt Mathematics

12-7 Lines of Best Fit

The table shows student enrollment at a college by year. The enrollment numbers are graphed below.

Year	2001	2002	2003	2004	2005	2006	2007	2008
Enrollment	950	995	1011	1020	1035			

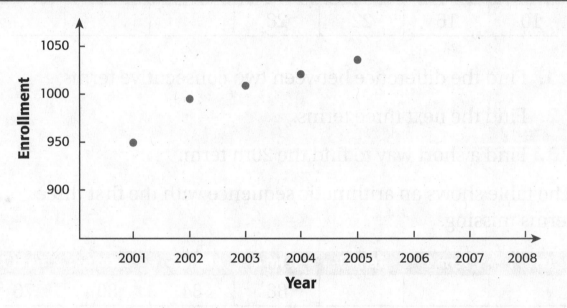

1. Find x_m, the mean of the *x*-values of the points on the graph.

2. Find y_m, the mean of the *y*-values of the points on the graph.

3. Plot (x_m, y_m). Lay the edge of a ruler on the graph through (x_m, y_m). Pivot the ruler around (x_m, y_m) and draw the line that you think is closest to the line of best fit.

Think and Discuss

4. **Predict** the enrollment for 2006, 2007, and 2008 by using the line of best fit.

Holt Mathematics

13-1 Terms of Arithmetic Sequences

In an *arithmetic sequence,* the difference between one term and the next is always the same.

The table shows the first four terms of an arithmetic sequence.

Term 1	Term 2	Term 3	Term 4	Term 5	Term 6	Term 7
10	16	22	28			

1. Find the difference between two consecutive terms.

2. Find the next three terms.

3. Find a short way to find the 20th term.

The table shows an arithmetic sequence with the first three terms missing.

Term 1	Term 2	Term 3	Term 4	Term 5	Term 6	Term 7
			88	84	80	76

4. Find the difference between two consecutive terms.

5. Find the first three terms.

Think and Discuss

6. **Discuss** the similarities and differences between the two sequences.

7. **Explain** your strategy for finding the 20th term in the first sequence.

Holt Mathematics

13-2 Terms of Geometric Sequences

In a *geometric sequence,* the ratio of one term to the next is always the same.

The table shows the first four terms of a geometric sequence.

Term 1	Term 2	Term 3	Term 4	Term 5	Term 6	Term 7
10	20	40	80			

1. Find the ratio between two consecutive terms.

2. Find the next three terms.

3. Find a short way to find the 20th term.

The table shows a geometric sequence with the first three terms missing.

Term 1	Term 2	Term 3	Term 4	Term 5	Term 6	Term 7
			150	75	37.5	18.75

4. Find the ratio between two consecutive terms.

5. Find the first three terms.

Think and Discuss

6. **Discuss** the similarities and differences between the two sequences.

7. **Explain** your strategy for finding the 20th term in the first sequence.

Holt Mathematics

13-3 Other Sequences

The table shows a sequence that is determined by the sum of consecutive odd numbers.

Term 1	Term 2	Term 3	Term 4	Term 5	Term 6	Term 7
1	1 + 3 = 4	1 + 3 + 5 = 9				

1. Find the next four terms.

2. Is the difference between consecutive terms a constant?

The table shows a sequence that is determined by the sum of consecutive even numbers.

Term 1	Term 2	Term 3	Term 4	Term 5	Term 6	Term 7
2	2 + 4 = 6	2 + 4 + 6 = 12				

3. Find the next four terms.

4. Is the difference between consecutive terms a constant?

Think and Discuss

5. **Describe** the sequence created by the differences of consecutive terms in the first table above.

6. **Describe** the sequence created by the differences of consecutive terms in the second table above.

Holt Mathematics

13-4 Linear Functions

Many everyday situations can be represented with *linear functions.*

For each situation, write a rule and complete the table.

1. A car beginning at time = 0 hours travels 60 miles per hour.

Input x (hr)	Rule $y =$ _____	Output y (mi)
0	$60 \cdot 0$	0
1	$60 \cdot 1$	60
2		120
3		
4		

2. A club that has $2000.00 in its treasury plans to spend $150.00 per week.

Input x (weeks)	Rule $y =$ _____	Output y ($)
0		2000
1		1850
2		1700
3		
4		

Think and Discuss

3. **Explain** how you determined the rules in Problems 1 and 2.

Holt Mathematics

13-5 Exponential Functions

Scientists use *exponential functions* to make predictions about populations. Consider the model for a deer population, which begins with 1000 deer in year 0.

1. Use a calculator to complete the table.

Input x (year)	Rule: $1000(1.1)^x$	Output y (number of deer)
0	$1000(1.1)^0$	1000
1	$1000(1.1)^1$	1100
2	$1000(1.1)^2$	
3		
4		
10		

2. Determine the year in which the deer population will be greater than 3000 for the first time.

Think and Discuss

3. Discuss the difference between a linear function and an exponential function.

4. Explain what percent increase per year in the deer population the exponential model shows.

Holt Mathematics

13-6 Quadratic Functions

The variable in a *quadratic function* is squared. The graph of $y = (x - 2)(x + 3)$ shows that the function has two x-intercepts, $x = -3$ and $x = 2$, and one y-intercept, $y = -6$.

   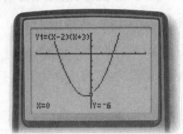

Graph each quadratic function on a graphing calculator. Use the given window settings to find the x- and y-intercepts.

1. $y = (x + 5)(x + 1)$

2. $y = (x - 4)(x - 2)$

Think and Discuss

3. Discuss the relationship between the function rule and the intercepts, the points where the graph crosses the x- and y-axes.

Holt Mathematics

13-7 Inverse Variation

In an *inverse variation,* when one quantity increases, the other decreases.

Water pressure is measured in pounds per square inch (psi). Water pressure decreases as the height at which each home is located increases.

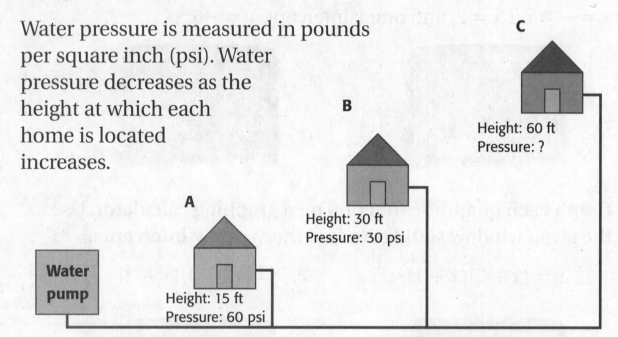

A
Height: 15 ft
Pressure: 60 psi

Water pump

B
Height: 30 ft
Pressure: 30 psi

C
Height: 60 ft
Pressure: ?

1. For house A and house B, multiply the water pressure by the height at which each house is located.

2. What do you notice about the products in Problem 1?

3. Predict the water pressure for house C.

Think and Discuss

4. **Explain** how you predicted the water pressure for house C.
5. **Give** another example of an inverse variation.

Holt Mathematics

EXPLORATION

14-1 Polynomials

An object is dropped from an initial height of 144 feet. The graph shows its height versus time.

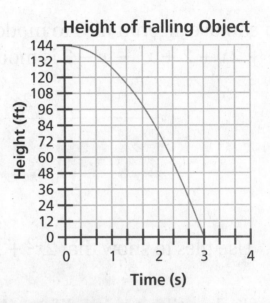

Height of Falling Object

Time (s)	Equation	Height (ft)
x	$y = 144 - 16x^2$	y
0	$y = 144 - 16(0)^2 = 144$	144
1		
2		
3		

1. What does the point (0, 144) represent?

2. When does the object reach the ground?

3. You can use the equation $y = 144 - 16x^2$ to model the object's fall. Complete the table and label the points on the graph.

Think and Discuss

4. **Explain** why the graph of a falling object is not a straight line.

Holt Mathematics

EXPLORATION

14-2 Simplifying Polynomials

You can use algebra tiles to model polynomials. The polynomial $2x^2 + 2x + 2 + x^2 - x - 3$ is modeled below.

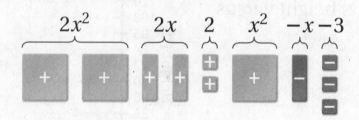

1. Use tiles to show that $2x^2 + 2x + 2 + x^2 - x - 3 = 3x^2 + x - 1$.

Use algebra tiles to simplify each expression.

2. $4x^2 - 2x - 5 - 3x^2 + x - 4$

3. $3x^2 - x + 1 - x^2 - x + 3$

Think and Discuss

4. **Explain** how you can use tiles to simplify polynomials.
5. **Explain** why you cannot simplify the polynomial $3x^2 + 4x - 9$.

Holt Mathematics

14-3 Adding Polynomials

You can use algebra tiles to model addition of polynomials. The addition problem $(2x^2 + 2x - 2) + (x^2 - 4x - 3)$ is modeled below.

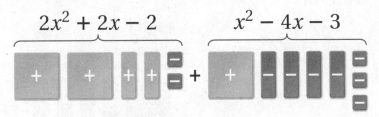

$$2x^2 + 2x - 2 \qquad\qquad x^2 - 4x - 3$$

1. Use tiles to show that
   $(2x^2 + 2x - 2) + (x^2 - 4x - 3) = 3x^2 - 2x - 5$.

You can use a graphing calculator to check. Enter the left side of the equation in **Y1** and the right side in **Y2,** and compare tables.

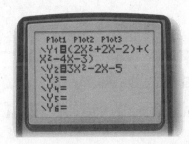 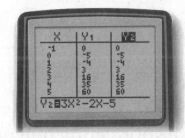

Use algebra tiles to add each pair of polynomials. Check your answers with a graphing calculator.

2. $(x^2 + 7x) + (3x^2 - 7x - 5)$

3. $(-3x^2 - x - 1) + (4x^2 - 3x - 2)$

Think and Discuss

4. **Explain** how you can use tiles to add polynomials.

Holt Mathematics

14-4 Subtracting Polynomials

You can use algebra tiles to find the opposite of a polynomial.
To do this, replace each tile with its opposite.

The opposite of

$$x^2 + 3x - 2$$

is

$$-x^2 - 3x + 2.$$

Use algebra tiles to find the opposite of each polynomial.

1. $4x^2 - 2x - 5$

2. $-x^2 - 7x + 3$

3. $2x^2 + 3x - 3$

Think and Discuss

4. **Explain** how you can use tiles to find opposites.

5. **Discuss** how opposites of polynomials might be useful if you were subtracting polynomials.

Holt Mathematics

14-5 Multiplying Polynomials by Monomials

You can use algebra tiles to model multiplication. The model shows that $2(-3) = -6$.

Find each product.

1. $2(2x - 3)$

2. $x(2x - 3)$

Use algebra tiles to model and find each product.

3. $2(3x + 2)$

4. $2x(3x - 1)$

Think and Discuss

5. Explain how area applies to modeling multiplication.

6. Write the factors and product modeled by the tiles shown.

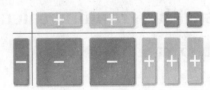

Holt Mathematics

14-6 Multiplying Binomials

You can use algebra tiles to model multiplication of binomials. The model shows the product of $(x + 2)(x + 4)$.

1. Use the model to find the product of $(x + 2)(x + 4)$.

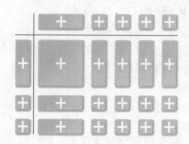

Use a graphing calculator to check your answer. Enter $(x + 2)(x + 4)$ for **Y1** and your answer for **Y2**, and compare tables.

Use algebra tiles to model and find each product. Check your answers with a graphing calculator.

2. $(x + 2)(x + 1)$

3. $(x + 2)(x + 3)$

Think and Discuss

4. Write the factors and product modeled by the tiles shown.

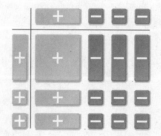

Holt Mathematics

Name _____ Date _____ Class _____

Problem Solving

1-1 *Variables and Expressions*

Write the correct answer.

1. If *l* is the length of a room and *w* is the width, then *lw* can be used to find the area of the room. Find the area of a room with *l* = 10 ft and *w* = 15 ft.

2. If *l* is the length of a room and *w* is the width, $2l + 2w$ can be used to find the perimeter of the room. Find the perimeter of a room with *l* = 12 ft and *w* = 16 ft.

3. Jaime earns 20% commission on her sales. If *s* is her total sales, then 0.2*s* can be used to find the amount she earns in commission. Find her commission if her sales are $1200.

4. If *p* is the regular hourly rate of pay, then 1.5*p* can be used to find the overtime rate of pay. Find the overtime rate of pay if the regular hourly rate of pay is $6.00 per hour.

Choose the letter for the best answer.

5. A plumber charges a fee of $75 per service call plus $15 per hour. If *h* is the number of hours the plumber works, then $75 + 15h$ can be used to find the total charges. Find the total charges if the plumber works 2.5 hours.

   A $37.50
   B $112.50
   C $225
   D $1127.50

6. Tickets to the movies cost $4 for students and $6 for adults. If *s* is the number of students and *a* is the number of adults, $4s + 6a$ can be used to find the cost of the tickets. Find the cost of the tickets for 3 students and 2 adults.

   F $15
   G $17
   H $24
   J $26

7. If *c* is the number of cricket chirps in a minute, then the expression $0.25c + 20$ can be used to estimate the temperature in degrees Farenheit. If there are 92 cricket chirps in a minute, find the temperature.

   A 43 degrees
   B 33 degrees
   C 102 degrees
   D 75 degrees

8. Flowers are sold in flats of 6 plants each. If *f* is the number of flats, then 6*f* can be used to find the number of flowers. Find the number of flowers in 18 flats.

   F 3 flowers
   G 108 flowers
   H 24 flowers
   J 12 flowers

PS1
Holt Mathematics

Name _____ Date _____ Class _____

Write the correct answer.

1. Morton bought 15 new books to add to his collection of books *b*. Write an algebraic expression to evaluate the total number of books in Morton's collection if he had 20 books in his collection.

2. Paul exercises *m* minutes per day 5 days a week. Write an algebraic expression to evaluate how many minutes Paul exercises each week if he exercises 45 minutes per day.

3. Helen bought 3 shirts that each cost *s* dollars. Write an algebraic expression to evaluate how much Helen spent in all if each shirt cost $22.

4. Claire makes *b* bracelets to divide evenly among four friends and herself. Write an algebraic expression to evaluate the number of bracelets each person will receive if Claire makes 15 bracelets.

Choose the letter for the best answer.

5. Jonas collects baseball cards. He has 245 cards in his collection. For his birthday, he received *r* more cards, then he gave his brother *g* cards. Which algebraic expression represents the total number of cards he now has in his collection?

   A $245 + r + g$

   B $245 - r - g$

   C $245 + r - g$

   D $r + g - 245$

6. Monique is saving money for a computer. She has *m* dollars saved. For her birthday, her dad doubled her money, but then she spent *s* dollars on a shirt. Which algebraic expression represents the amount of money she has now saved for her computer?

   F $m + 2 - s$

   G $2m - s$

   H $2m + s$

   J $m + 2s$

7. Which algebraic expression represents the number of years in *m* months?

   A $12m$

   B $\dfrac{m}{12}$

   C $12 + m$

   D $12 - m$

8. Which algebraic expression represents how many minutes are in *h* hours?

   F $60h$

   G $\dfrac{h}{60}$

   H $h + 60$

   J $h - 60$

Holt Mathematics

Name _____ Date _____ Class _____

Problem Solving
Integers and Absolute Value

Write the correct answer.

1. In Africa, Lake Asal reaches a depth of −153 meters. In Asia, the Dead Sea reaches a depth of −408 meters. Which reaches a greater depth, Lake Asal or the Dead Sea?

_____

2. Jeremy's scores for four golf games are: −1, 2, −3, and 1. Order his golf scores from least to greatest.

_____

3. The lowest point in North America is Death Valley with an elevation of −282 feet. South America's lowest point is the Valdes Peninsula with an elevation of −131 feet. Which continent has the lowest point?

_____

4. Two undersea cameras are taking time lapse photos in a coral reef. The first camera is mounted at −45 feet. The second camera is mounted at −25. Which camera is closer to the surface?

_____

**Use the table to answer Exercises 5–7.
Choose the letter of the best answer.**

5. Which state had the coldest temperature?

   A Alabama   **C** Massachusetts
   B Indiana   **D** Texas

6. Which is the greatest temperature listed?

   F −27°F   **H** −35°F
   G −36°F   **J** −23°F

7. The lowest temperature recorded in Connecticut was between the lowest temperatures recorded in Alabama and Massachusetts. Which could be the lowest temperature recorded in Connecticut?

   A −40°F   **C** −32°F
   B −37°F   **D** −40°F

State Low Temperature Records

State	Temperature (°F)
Alabama	−27
Indiana	−36
Massachusetts	−35
Texas	−23

Holt Mathematics

LESSON 1-4 Problem Solving
Adding Integers

Use the following information for Exercises 1–3. In golf, par 73 means that a golfer should take 73 strokes to finish 18 holes. A score of 68 is 5 under par, or −5. A score of 77 is 4 over par, or +4.

1. Use integers to write Tiger Woods's score for each round as over or under par.

_____

2. Add the integers to find Tiger Woods's overall score.

_____

3. Was Tiger Woods's overall score over or under par?

_____

_____

**Tiger Woods's Scores
Mercedes Championship
January 6, 2002
Par 73 course**

Round	Score
1	68
2	74
3	74
4	65

Choose the letter for the best answer.

4. At 9:00 A.M., the temperature was −15°. An hour later, the temperature had risen 7°. What is the temperature now?

  A −22°   C −8°

  B 8°   D 22°

5. Sandra is reviewing her savings account statement. She withdrew amounts of $35, $20, and $15. She deposited $65. If her starting balance was $657, find the new balance.

  F $652   H $662

  G $522   J $507

6. During a possession in a football game, the Vikings gained 22 yards, lost 15 yards, gained 3 yards, gained 20 yards and lost 5 yards. At the end of the possession, how many yards had they lost or gained?

  A gained 43 yards

  B lost 43 yards

  C lost 25 yards

  D gained 25 yards

7. A submarine is cruising at 40 m below sea level. The submarine ascends 18 m. What is the submarine's new location?

  F 58 m below sea level

  G 22 m below sea level

  H 18 m below sea level

  J 12 m below sea level

Holt Mathematics

Name _____ Date _____ Class _____

Write the correct answer.

1. In Fairbanks, Alaska, the average January temperature is −13°F, while the average April temperature is 30°F. What is the difference between the average temperatures?

2. The highest point in North America is Mt. McKinley, Alaska, at 20,320 ft above sea level. The lowest point is Death Valley, California, at 282 ft below sea level. What is the difference in elevations?

3. The temperature fell from 44°F to −56°F in 24 hours in Browning, Montana, on January 23–24, 1916. By how many degrees did the temperature change?

4. The boiling point of chlorine is −102°C, while the melting point is −34°C. What is the difference between the melting and boiling points of chlorine?

Use the table below to answer Exercises 5–7. The table shows the first and fifth place finishers in a golf tournament. In golf, the winner has the lowest total for all five rounds. Choose the letter for the best answer.

5. By how many points did Mickelson beat Kelly in Round 2?

   A 2   **C** 5
   B 3   **D** 8

6. By how many points did Kelly beat Mickelson in Round 3?

   F 2   **H** 5
   G 3   **J** 9

7. Who won the Bob Hope Chrysler Classic and how many points difference was there between first and fifth place?

   A Kelly; 4   **C** Kelly; 3
   B Mickelson; 4   **D** Mickelson; 3

Bob Hope Chrysler Classic
January 20, 2002

Round	J. Kelly	P. Mickelson
1	−8	−8
2	−3	−5
3	−7	−2
4	−4	−7
5	−5	−8

Holt Mathematics

LESSON 1-6 Problem Solving
Multiplying and Dividing Integers

Write the correct answer.

1. A submersible started at the surface of the water and was moving down at −12 meters per minute toward the ocean floor. The submersible traveled at this rate for 32 minutes before coming to rest on the ocean floor. What is the depth of the ocean floor?

_____

2. For the first week in January, the daily high temperatures in Bismarck, North Dakota, were 7°F, −10°F, −10°F, −7°F, 8°F, 12°F, and 14°F. What was the average daily high temperature for the week?

_____

3. Sally went golfing and recorded her scores as −2 on the first hole, −2 on the second hole, and 1 on the third hole. What is her average for the first three holes?

_____

4. The ocean floor is at −96 m. Tom has reached −15 m. If he continues to move down at −3 m per minute, how far will he be from the ocean floor after 7 minutes?

_____

Use the table below to answer Exercises 5–7. Choose the letter for the best answer.

5. What is the caloric impact of 2 hours of in-line skating?

   A −477 Cal   C −583 Cal

   B −479 Cal   D −954 Cal

6. What is the caloric impact of eating a hamburger and then playing Frisbee for 3 hours?

   F 220 Cal   H 190 Cal

   G −190 Cal   J −220 Cal

7. Tim plays basketball for 1 hour, skates for 5 hours, and plays Frisbee for 4 hours. What is the average amount of calories Tim burns per hour?

   A −375 Cal   C −545 Cal

   B −1250 Cal   D −409 Cal

Calories Consumed or Burned

Food or Exercise	Calories
Apple	125
Pepperoni pizza (slice)	181
Hamburger	425
Basketball (1hr)	−545
In-line skating (1 hr)	−477
Frisbee (1 hr)	−205

Holt Mathematics

Name _____ Date _____ Class _____

Problem Solving

Solving Equations by Adding or Subtracting

Write the correct answer.

1. The 1954 elevation of Mt. Everest was 29,028 ft. In 1999, that elevation was revised to be 29,035 ft. Write an equation to find the change c in elevation of Mt. Everest.

2. The difference between the boiling and melting points of fluorine is 32°C. If the boiling point of fluorine is -188°C, write an equation and solve to find the melting point m of fluorine.

3. Lisa sold her old bike for $140 less than she paid for it. She sold the bike for $85. Write and solve an equation to find how much Lisa paid for her bike.

4. The average January temperature in Fairbanks, Alaska, is -13°F. The April average is 43°F higher than the January average. Write an equation to find the average April temperature.

Choose the letter for the best answer.

5. A survey found that female teens watched 3 hours of TV per week less than male teens. The female teens reported watching an average of 18 hours of TV. Find the number of hours h the male teens watched.

   A $h = 6$ **C** $h = 18$

   B $h = 15$ **D** $h = 21$

6. It costs about $125 more per year to feed a hamster than it does to feed a bird. If it costs $256 per year to feed a hamster, find the cost c to feed a bird.

   F $c = \$131$ **H** $c = \$256$

   G $c = \$125$ **J** $c = \$381$

7. Naples, Florida, is the second fastest growing U.S. metropolitan area. From 1990 to 2000, the population increased by 99,278. If the 2000 population was 251,377, find the population p in 1990.

   A $p = 253,377$ **C** $p = 249,377$

   B $p = 350,655$ **D** $p = 152,099$

8. In 1940, the life expectancy for a female was 65 years. In 1999, the life expectancy for a female was 79 years. Find the increase in the life expectancy for females.

   F 14 yrs   **H** -14 yrs

   G 1.2 yrs   **J** 144 yrs

Holt Mathematics

LESSON 1-8 Problem Solving
Solving Equations by Multiplying or Dividing

Write the correct answer.

1. Brett is preparing to participate in a 250-kilometer bike race. He rides a course near his house that is 2 km long. Write an equation to determine how many laps he must ride to equal the distance of the race.

2. The average life span of a duck is 10 years, which is one year longer than three times the average life span of a guinea pig. Write and solve an equation to determine the lifespan of a guinea pig.

3. The speed of a house mouse is one-fourth that of a giraffe. If a house mouse can travel at 8 mi/h, what is the speed of a giraffe? Write an equation and solve.

4. In 2005, the movie with the highest box office sales was *Titanic,* which made about 3 times the box office sales of *Charlie and the Chocolate Factory*. If *Titanic* made about $600 million, about how much did *Charlie and the Chocolate Factory* make? Write an equation and solve.

Choose the letter for the best answer.

5. Farmland is often measured in acres. A farm that is 1920 acres covers 3 square miles. Find the number of acres *a* in one square mile.

   A 9 acres   C 640 acres

   B 213 acres   D 4860 acres

6. When Maria doubles a recipe, she uses 8 cups of flour. How many cups of flour are in the original recipe?

   F 2 cups   H 8 cups

   G 4 cups   J 16 cups

7. The depth of water is often measured in fathoms. A fathom is six feet. If the maximum depth of the Gulf of Mexico is 2395 fathoms, what is the maximum depth in feet?

   A 14,370 ft   C 29,250 ft

   B 98,867 ft   D 175,464 ft

8. Four times as many pet birds have lived in the White House as pet goats. Sixteen pet birds have lived in the White House. How many pet goats have there been?

   F 4   H 12

   G 20   J 64

Holt Mathematics

LESSON 1-9

Problem Solving
Introduction to Inequalities

Use the table.

1. Write an inequality that compares the population p of Los Angeles to the population of New York.

_____

2. Write an inequality that compares the population p of Los Angeles to the population of Chicago.

_____

Top 3 U.S. Cities by Population 2000

Rank	City	Population
1	New York	8,008,278
2	Los Angeles	p
3	Chicago	2,896,016

Write the correct answer.

3. Paul wants to ride his bike at least 30 miles this week to train for a race. He has already ridden 18 miles. How many more miles should Paul ride this week?

_____

4. To avoid a service charge, Jose must keep more than $500 in his account. His current balance is $536, but he plans to write a check for $157. Find the amount of the deposit d Jose must make to avoid a service charge.

_____

Choose the letter for the best answer.

5. Mia wants to spend no more than $10 on an ad in the paper. The first 10 words cost $3. Find the amount of money m she has left to spend on the ad.

   A $m \geq 7$ **C** $m \leq 7$

   B $m \leq 13$ **D** $m \geq 13$

6. An auto shop estimates parts and labor for a repair will cost less than $200. Parts will cost $59. Find the maximum cost c of the labor.

   F $c < \$141$ **H** $c > \$141$

   G $c < \$259$ **J** $c > \$259$

7. To advance to the next level of a competition, Rachel must earn at least 180 points. She has already earned 145 points. Find the number of points p she needs to advance to the next level of the competition.

   A $p \leq 35$ **C** $p \geq 35$

   B $p \leq 325$ **D** $p \geq 325$

8. The Conway's hiked more than 25 miles on their backpacking trip. If they hiked 8 miles on their last day, find how many miles m they hiked on the rest of the trip.

   F $m > 17$ **H** $m < 17$

   G $m > 33$ **J** $m < 33$

Holt Mathematics

LESSON 2-1	**Problem Solving**
	Rational Numbers

Write the correct answer.

1. Fill in the table below which shows the sizes of drill bits in a set.

2. Do the drill bit sizes convert to repeating or terminating decimals?

_____

13-Piece Drill Bit Set

Fraction	Decimal	Fraction	Decimal	Fraction	Decimal
$\frac{1}{4}$"		$\frac{11}{64}$"		$\frac{3}{32}$"	
$\frac{15}{64}$"		$\frac{5}{32}$"		$\frac{5}{64}$"	
$\frac{7}{32}$"		$\frac{9}{64}$"		$\frac{1}{16}$"	
$\frac{13}{64}$"		$\frac{1}{8}$"			
$\frac{3}{16}$"		$\frac{7}{64}$"			

Use the table at the right that lists the world's smallest nations. Choose the letter for the best answer.

3. What is the area of Vatican City expressed as a fraction in simplest form?

   A $\frac{8}{50}$ C $\frac{17}{1000}$

   B $\frac{4}{25}$ D $\frac{17}{100}$

World's Smallest Nations

Nation	Area (square miles)
Vatican City	0.17
Monaco	0.75
Nauru	8.2

4. What is the area of Monaco expressed as a fraction in simplest form?

   F $\frac{75}{100}$ H $\frac{3}{4}$

   G $\frac{15}{20}$ J $\frac{2}{3}$

5. What is the area of Nauru expressed as a mixed number?

   A $8\frac{1}{50}$ C $8\frac{2}{100}$

   B $8\frac{2}{50}$ D $8\frac{1}{5}$

6. The average annual precipitation in Miami, FL is 57.55 inches. Express 57.55 as a mixed number.

   F $57\frac{11}{20}$ H $57\frac{5}{100}$

   G $57\frac{55}{1000}$ J $57\frac{1}{20}$

7. The average annual precipitation in Norfolk, VA is 45.22 inches. Express 45.22 as a mixed number.

   A $45\frac{11}{50}$ C $45\frac{11}{20}$

   B $45\frac{22}{1000}$ D $45\frac{1}{5}$

Holt Mathematics

Problem Solving

LESSON 2-2 *Comparing and Ordering Rational Numbers*

Write the correct answer.

1. Carl Lewis won the gold medal in the long jump in four consecutive Summer Olympic games. He jumped 8.54 meters in 1984, 8.72 meters in 1988, 8.67 meters in 1992, 8.5 meters in 1996. Order the length of his winning jumps from least to greatest.

3. The depth of a lake is measured at three different points. Point A is −15.8 meters, Point B is −17.3 meters, and Point C is −16.9 meters. Which point has the greatest depth?

2. Scientists aboard a submarine are gathering data at an elevation of $-42\frac{1}{2}$ feet. Scientists aboard a submersible are taking photographs at an elevation of $-45\frac{1}{3}$ feet. Which scientists are closer to the surface of the ocean?

4. At a swimming meet, Gail's time in her first heat was $42\frac{3}{8}$ seconds. Her time in the second heat was 42.25 seconds. Which heat did she swim faster?

The table shows the top times in a 5 K race. Choose the letter of the best answer.

5. Who had the fastest time in the race?
   A Marshall
   B Renzo
   C Dan
   D Aaron

Name	Time (minutes)
Marshall	18.09
Renzo	17.38
Dan	17.9
Aaron	18.61

6. Which is the slowest time in the table?
   F 18.09 minutes
   G 17.38
   H 17.9 minutes
   J 18.61 minutes

7. Aaron's time in a previous race was less than his time in this race but greater than Marshall's time in this race. How fast could Aaron have run in the previous race?
   A 19.24 min   C 18.35 min
   B 18.7 min   D 18.05 mi

Holt Mathematics

Name _____ Date _____ Class _____

Problem Solving
Adding and Subtracting Rational Numbers

Write the correct answer.

1. In 2004, Yuliya Nesterenko of Belarus won the Olympic Gold in the 100-m dash with a time of 10.93 seconds. In 2000, American Marion Jones won the 100-m dash with a time of 10.75 seconds. How many seconds faster did Marion Jones run the 100-m dash?

2. The snowfall in Rochester, NY in the winter of 1999–2000 was 91.5 inches. Normal snowfall is about 76 inches per winter. How much more snow fell in the winter of 1999–2000 than is normal?

3. In a survey, $\frac{76}{100}$ people indicated that they check their e-mail daily, while $\frac{23}{100}$ check their e-mail weekly, and $\frac{1}{100}$ check their e-mail less than once a week. What fraction of people check their e-mail at least once a week?

4. To make a small amount of play dough, you can mix the following ingredients: 1 cup of flour, $\frac{1}{2}$ cup of salt and $\frac{1}{2}$ cup of water. What is the total amount of ingredients added to make the play dough?

Choose the letter for the best answer.

5. How much more expensive is it to buy a ticket in Boston than in Minnesota?

   A $20.95   C $5.40

   B $55.19   D $26.35

Baseball Ticket Prices

Location	Average Price
Minnesota	$14.42
League Average	$19.82
Boston	$40.77

6. How much more expensive is it to buy a ticket in Boston than the league average?

   F $60.59

   G $20.95

   H $5.40

   J $26.35

7. What is the total cost of a ticket in Boston and a ticket in Minnesota?

   A $55.19

   B $34.24

   C $60.59

   D $54.19

Holt Mathematics

Name _____ Date _____ Class _____

LESSON 2-4 **Problem Solving**
Multiplying Rational Numbers

Use the table at the right.

1. What was the average number of births per minute in 2001?

   _____

Average World Births and Deaths per Second in 2001	
Births	$4\frac{1}{5}$
Deaths	1.7

2. What was the average number of deaths per hour in 2001?

   _____

3. What was the average number of births per day in 2001?

   _____

4. What was the average number of births in $\frac{1}{2}$ of a second in 2001?

   _____

5. What was the average number of births in $\frac{1}{4}$ of a second in 2001?

   _____

Use the table below. During exercise, the target heart rate is 0.5–0.75 of the maximum heart rate. Choose the letter for the best answer.

6. What is the target heart rate range for a 14 year old?

   A 7–10.5

   B 103–154.5

   C 145–166

   D 206–255

Age	Maximum Heart Rate
13	207
14	206
15	205
20	200
25	195

Source: American Heart Association

7. What is the target heart rate range for a 20 year old?

   F 100–150

   G 125–175

   H 150–200

   J 200–250

8. What is the target heart rate range for a 25 year old?

   A 25–75

   B 85–125

   C 97.5–146.25

   D 195–250

Holt Mathematics

Problem Solving

Dividing Rational Numbers

Use the table at the right that shows the maximum speed over a quarter mile of different animals. Find the time is takes each animal to travel one-quarter mile at top speed. Round to the nearest thousandth.

1. Quarter horse

_____

2. Greyhound

_____

3. Human

_____

4. Giant tortoise

_____

Maximum Speeds of Animals

Animal	Speed (mph)
Quarter Horse	47.50
Greyhound	39.35
Human	27.89
Giant Tortoise	0.17
Three-toed sloth	0.15

5. Three-toed sloth

_____

Choose the letter for the best answer.

6. A piece of ribbon is $1\frac{7}{8}$ inches long. If the ribbon is going to be divided into 15 pieces, how long should each piece be?

A $\frac{1}{8}$ in.

B $\frac{1}{15}$ in.

C $\frac{2}{3}$ in.

D $28\frac{1}{8}$ in.

7. The recorded rainfall for each day of a week was 0 in., $\frac{1}{4}$ in., $\frac{3}{4}$ in., 1 in., 0 in., $1\frac{1}{4}$ in., $1\frac{1}{4}$ in. What was the average rainfall per day?

F $\frac{9}{10}$ in.

G $\frac{9}{14}$ in.

H $\frac{7}{8}$ in.

J $4\frac{1}{2}$ in.

8. A drill bit that is $\frac{7}{32}$ in. means that the hole the bit makes has a diameter of $\frac{7}{32}$ in. Since the radius is half of the diameter, what is the radius of a hole drilled by a $\frac{7}{32}$ in. bit?

A $\frac{14}{32}$ in.   C $\frac{9}{16}$ in.

B $\frac{7}{32}$ in.   D $\frac{7}{64}$ in.

9. A serving of a certain kind of cereal is $\frac{2}{3}$ cup. There are 12 cups of cereal in the box. How many servings of cereal are in the box?

F 18

G 15

H 8

J 6

Holt Mathematics

Name _____ Date _____ Class _____

LESSON 2-6 Problem Solving
Adding and Subtracting with Unlike Denominators

Write the correct answer.

1. Nick Hysong of the United States won the Olympic gold medal in the pole vault in 2000 with a jump of 19 ft $4\frac{1}{4}$ inches, or $232\frac{1}{4}$ inches. In 1900, Irving Baxter of the United States won the pole vault with a jump of 10 ft $9\frac{7}{8}$ inches, or $129\frac{7}{8}$ inches. How much higher did Hysong vault than Baxter?

2. In the 2000 Summer Olympics, Ivan Pedroso of Cuba won the Long jump with a jump of 28 ft $\frac{3}{4}$ inches, or $336\frac{3}{4}$ inches. Alvin Kraenzlein of the Unites States won the long jump in 1900 with a jump of 23 ft $6\frac{7}{8}$ inches, or $282\frac{7}{8}$ inches. How much farther did Pedroso jump than Kraenzlein?

3. A recipe calls for $\frac{1}{8}$ cup of sugar and $\frac{3}{4}$ cup of brown sugar. How much total sugar is added to the recipe?

4. The average snowfall in Norfolk, VA for January is $2\frac{3}{5}$ inches, February $2\frac{9}{10}$ inches, March 1 inch, and December $\frac{9}{10}$ inches. If these are the only months it typically snows, what is the average snowfall per year?

Use the table at the right that shows the average snowfall per month in Vail, Colorado.

5. What is the average annual snowfall in Vail, Colorado?

   A $15\frac{13}{20}$ in.   C $187\frac{1}{10}$ in.

   B 153 in.   D $187\frac{4}{5}$ in.

6. The peak of the skiing season is from December through March. What is the average snowfall for this period?

   F $30\frac{19}{20}$ in.   H $123\frac{4}{5}$ in.

   G $123\frac{3}{5}$ in.   J 127 in.

Average Snowfall in Vail, CO

Month	Snowfall (in.)	Month	Snowfall (in.)
Jan	$36\frac{7}{10}$	July	0
Feb	$35\frac{7}{10}$	August	0
March	$25\frac{2}{5}$	Sept	1
April	$21\frac{1}{5}$	Oct	$7\frac{4}{5}$
May	4	Nov	$29\frac{7}{10}$
June	$\frac{3}{10}$	Dec	26

Holt Mathematics

Problem Solving
Solving Equations with Rational Numbers

Write the correct answer.

1. In the last 150 years, the average height of people in industrialized nations has increased by $\frac{1}{3}$ foot. Today, American men have an average height of $5\frac{7}{12}$ feet. What was the average height of American men 150 years ago?

2. Jaime has a length of ribbon that is $23\frac{1}{2}$ in. long. If she plans to cut the ribbon into pieces that are $\frac{3}{4}$ in. long, into how many pieces can she cut the ribbon? (She cannot use partial pieces.)

3. Todd's restaurant bill for dinner was $15.55. After he left a tip, he spent a total of $18.00 on dinner. How much money did Todd leave for a tip?

4. The difference between the boiling point and melting point of Hydrogen is 6.47°C. The melting point of Hydrogen is −259.34°C. What is the boiling point of Hydrogen?

Choose the letter for the best answer.

5. Justin Gatlin won the Olympic gold in the 100-m dash in 2004 with a time of 9.85 seconds. His time was 0.95 seconds faster than Francis Jarvis who won the 100-m dash in 1900. What was Jarvis' time in 1900?

   A 8.95 seconds

   B 10.65 seconds

   C 10.80 seconds

   D 11.20 seconds

6. The balance in Susan's checking account was $245.35. After the bank deposited interest into the account, her balance went to $248.02. How much interest did the bank pay Susan?

   F $1.01

   G $2.67

   H $3.95

   J $493.37

7. After a morning shower, there was $\frac{17}{100}$ in. of rain in the rain gauge. It rained again an hour later and the rain gauge showed $\frac{1}{4}$ in. of rain. How much did it rain the second time?

   A $\frac{2}{25}$ in.   C $\frac{21}{50}$ in.

   B $\frac{1}{6}$ in.   D $\frac{3}{8}$ in.

8. Two-third of John's savings account is being saved for his college education. If $2500 of his savings is for his college education, how much money in total is in his savings account?

   F $1666.67   H $4250.83

   G $3750   J $5000

Holt Mathematics

Name _____ Date _____ Class _____

Problem Solving

Solving Two-Step Equations

The chart below describes three different long distance calling plans. Jamie has budgeted $20 per month for long distance calls. Write the correct answer.

1. How many minutes will Jamie be able to use per month with plan A? Round to the nearest minute.

   _____

Plan	Monthly Access Fee	Charge per minute
A	$3.95	$0.08
B	$8.95	$0.06
C	$0	$0.10

2. How many minutes will Jamie be able to use per month with plan B? Round to the nearest minute.

   _____

3. How many minutes will Jamie be able to use per month with plan C? Round to the nearest minute.

   _____

4. Which plan is the best deal for Jamie's budget?

   _____

5. Nolan has budgeted $50 per month for long distance. Which plan is the best deal for Nolan's budget?

   _____

The table describes four different car loans that Susana can get to finance her new car. The total column gives the amount she will end up paying for the car including the down payment and the payments with interest. Choose the letter for the best answer.

6. How much will Susana pay each month with loan A?

   A $252.04   C $330.35

   B $297.02   D $353.68

Loan	Down Payment	Number of Months	Total
A	$2000	60	$19,821.20
B	$1000	48	$19,390.72
C	$0	60	$20,197.20

7. How much will Susana pay each month with loan B?

   F $300.85   H $323.17

   G $306.50   J $383.14

8. How much will Susana pay each month with loan C?

   A $336.62   C $369.95

   B $352.28   D $420.78

9. Which loan will give Susana the smallest monthly payment?

   F Loan A   H Loan C

   G Loan B   J They are equal

Holt Mathematics

Name _____ Date _____ Class _____

Problem Solving
Ordered Pairs

Use the table at the right for Exercises 1–2.

1. Write the ordered pair that shows the average miles per gallon in 1990.

   _____

2. The data can be approximated by the equation $m = 0.30887x - 595$ where m is the average miles per gallon and x is the year. Use the equation to find an ordered pair (x, m) that shows the estimated miles per gallon in the year 2020.

   _____

Average Miles per Gallon

Year	Miles per Gallon
1970	13.5
1980	15.9
1990	20.2
1995	21.1
1996	21.2
1997	21.5

For Exercises 3–4 use the equation $F = 1.8C + 32$, which relates Fahrenheit temperatures F to Celsius temperatures C.

3. Write ordered pair (C, F) that shows the Celsius equivalent of 86°F.

   _____

4. Write ordered pair (C, F) that shows the Fahrenheit equivalent of 22°C.

   _____

Choose the letter for the best answer.

5. A taxi charges a $2.50 flat fee plus $0.30 per mile. Use an equation for taxi fare t in terms of miles m. Which ordered pair (m, t) shows the taxi fare for a 23-mile cab ride?

   A (23, 6.90)   **C** (23, 9.40)

   B (23, 18.50)   **D** (23, 64.40)

6. The perimeter p of a square is four times the length of a side s, or $p = 4s$. Which ordered pair (s, p) shows the perimeter for a square that has sides that are 5 in.?

   F (5, 1.25)   **H** (5, 9)

   G (5, 20)   **J** (5, 25)

7. Maria pays a monthly fee of $3.95 plus $0.10 per minute for long distance calls. Use an equation for the phone bill p in terms of the number of minutes m. Which ordered pair (m, p) shows the phone bill for 120 minutes?

   A (120, 15.95)   **C** (120, 28.30)

   B (120, 474.10) **D** (120, 486.00)

8. Tickets to a baseball game cost $12 each, plus $2 each for transportation. Use an equation for the cost c of going to the game in terms of the number of people p. Which ordered pair (p, c) shows the cost for 6 people?

   F (6, 74)   **H** (6, 84)

   G (6, 96)   **J** (6, 102)

Holt Mathematics

Name _____ Date _____ Class _____

LESSON **Problem Solving**
3-2 *Graphing on a Coordinate Plane*

Complete the table of ordered pairs. Graph each ordered pair.
Draw a line through the points. Answer the question.

1. John earns $150 per week plus 5% of his computer software sales. John's weekly pay *y* in terms of his sales *x* is $y = 150 + 0.05x$. Complete the table. How much does John get paid for $200 in sales?

_____

2. Margarite starts out with $100. Each week, she spends $6 to go to the movies. The amount of money *y* Margarite has left each week *x*, is $y = 100 - 6x$. How much money does she have left after 11 weeks?

_____

x	y	(x, y)
0		
10		
25		
50		
100		

x	y	(x, y)
0		
1		
2		
3		
4		

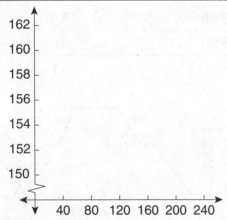

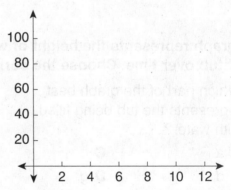

The graph at the right represents the miles traveled *y* in *x* hours. Use the graph to choose the best letter.

3. Which of the ordered pairs below represents a solution?

   A (1, 65)   **C** (5, 120)

   B (2, 70)   **D** (7, 300)

4. The graph represents a car traveling how fast?

   F 60 mi/h   **H** 70 mi/h

   G 65 mi/h   **J** 75 mi/h

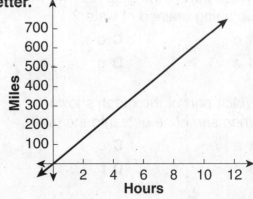

Holt Mathematics

Name _____ Date _____ Class _____

Problem Solving
Interpreting Graphs and Table

Tell which table corresponds to each situation.

1. Ryan walks for several blocks, and then he begins to run. After running for 10 minutes, he walks for several blocks and then stops.

_____

2. Susanna starts running. After 10 minutes, she sees a friend and stops to talk. When she leaves her friend, she runs home and stops.

_____

3. Mark stands on the porch and talks to a friend. Then he starts walking home. Part way home he decides to run the rest of the way, and he doesn't stop until he gets home.

_____

Table 1

Time	Speed (mi/h)
8:00	0
8:10	3
8:20	7.5
8:30	0

Table 2

Time	Speed (mi/h)
8:00	3
8:10	7.5
8:20	3
8:30	0

Table 3

Time	Speed (mi/h)
8:00	7.5
8:10	0
8:20	7.5
8:30	0

The graph represents the height of water in a bathtub over time. Choose the correct letter.

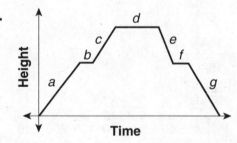

4. Which part of the graph best represents the tub being filled with water?

  A a   **C** c
  B d   **D** g

5. Which part of the graph shows the tub being drained of water?

  A c   **C** d
  B e   **D** g

6. Which part of the graph shows someone soaking in the tub?

  F b   **H** d
  G e   **J** f

7. Which part of the graph shows when someone gets into the tub?

  A a   **C** c
  B e   **D** f

8. Which parts of the graph show when the water level is not changing in the tub?

  F a, b, c   **H** b, d, g
  G b, d, f   **J** c, e, f

Holt Mathematics

Name _____ Date _____ Class _____

A cyclist rides at an average speed of 20 miles per hour. The
equation $y = 20x$ shows the distance, y, the cyclist travels in x hours.

1. Make a table for the equation and
graph the equation at the right.

x	$20x$	y
0		
1		
2		
3		

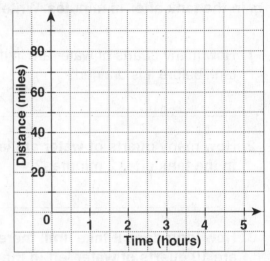

2. Is the relationship between the time
and the distance the cyclist rides a
function?

_____

3. If the cyclist continues to ride at the
same rate, about how far will the
cyclist ride in 4 hours?

_____

4. About how far does the cyclist ride in
1.5 hours?

_____

5. If the cyclist has ridden 50 miles, about
how long has the cyclist been riding?

_____

**The cost of renting a jet-ski at a lake is represented by the
equation $f(x) = 25x + 100$ where x is the number of hours and
$f(x)$ is the cost including an hourly rate and a deposit. Choose
the letter for the best answer.**

6. What is the domain of the function?

  A $x < 0$ **C** $x > 25$

  B $x > 0$ **D** $x < 100$

7. What is the range of the function?

  F $f(x) > 0$ **H** $f(x) < 25$

  G $f(x) < 0$ **J** $f(x) > 100$

8. How much does it cost to rent the
jet-ski for 5 hours?

  A $125   **C** $385

  B $225   **D** $525

9. If the cost to rent the jet-ski is $300,
for how many hours is the jet-ski
rented?

  F 6 hours   **H** 12 hours

  G 8 hours   **J** 16 hours

Holt Mathematics

Name _____ Date _____ Class _____

Problem Solving
Equations, Tables, and Graphs

Use the graph to answer Exercises 1–4. An aquarium tank is being drained. The graph shows the number of quarts of water, q, in the tank after m minutes. Write the correct answer.

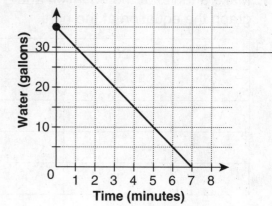

1. How many quarts of water are in the tank before it is drained?

   _____

2. How many quarts of water are left in the tank after 2 minutes?

   _____

3. How long does it take until there are 10 quarts of water left in the tank?

Use the graph to answer Exercises 5–7. The graph shows the distance, d, a hiker can hike in h hours. Choose the letter of the best answer.

5. How far can the hiker hike in 4 hours?

   A $1\frac{1}{3}$ mi   C 8 mi

   B 4 mi   D 12 mi

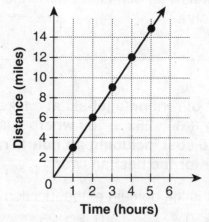

6. How long does it take the hiker to hike 6 miles?

   F 2 h H 4 h
   G 3 h J 18 h

7. Which equation represents the graph?

   A $d = 3h$ C $d = h + 3$

   B $d = \frac{1}{3}h$ D $d = h - 3$

Holt Mathematics

Name _____ Date _____ Class _____

Problem Solving
LESSON 3-6 *Arithmetic Sequences*

Write the correct answer.

1. An English teacher gives her class 6 vocabulary words on Monday. Each day for the rest of the week she adds 3 more vocabulary words to the list. How many words are on the list on Friday?

_____

2. A cab ride costs $1.50 plus $2.00 for each mile. What is the total cost of a 5-mile cab ride?

_____

3. Rosie ran 8 laps around the track. Each week after that she ran 3 more laps than the week before. How many laps will she run around the track in the sixth week?

4. Lee has saved $85. Each week he uses his savings to buy a CD for $9. How much money will he have left after the fourth week?

Use the table to answer Exercises 5–7. The table shows the number of seats in each row of a theater. Choose the letter of the best answer.

5. The number of seats is an arithmetic sequence. What is the common difference?

   A 6   C 9

   B 8   D 35

Row	Number of Seats
1	35
2	43
3	51
4	59
5	67

6. If the sequence continues, how many seats will be in the next row?

   F 68   H 75

   G 73   J 76

8. A class is taking a field trip to the zoo. Admission for the class costs $50 plus $2 for each student to visit the special exhibits. Which function best describes the total cost for *n* students?

   F $y = 50n - 2$
   G $y = 50n + 2$
   H $y = 50 - 2n$
   J $y = 50 + 2n$

7. If the sequence continues, how many seats will be in the tenth row?

   A 80   C 134

   B 107   D 147

Holt Mathematics

Name _____ Date _____ Class _____

Problem Solving
Exponents

Write the correct answer.

1. The formula for the volume of a cube is $V = e^3$ where e is the length of a side of the cube. Find the volume of a cube with side length 6 cm.

2. The distance in feet traveled by a falling object is given by the formula $d = 16t^2$ where t is the time in seconds. Find the distance an object falls in 4 seconds.

3. The surface area of a cube can be found using the formula $S = 6e^2$ where e is the length of a side of the cube. Find the surface area of a cube with side length 6 cm.

4. John's father offers to pay him 1 cent for doing the dishes the first night, 2 cents for doing the dishes the second, 4 cents for the third, and so on, doubling each night. Write an expression using exponents for the amount John will get paid on the tenth night.

Use the table below for Exercises 5–7, which shows the number of e-mails forwarded at each level if each person continues a chain by forwarding an e-mail to 10 friends. Choose the letter for the best answer.

5. How many e-mails were forwarded at level 5 alone?

   A 5^{10} C 2^{10}

   B 2^5 D 10^5

6. How many e-mails were forwarded at level 6 alone?

   F 100,000   H 10,000,000

   G 1,000,000   J 100,000,000

7. Forwarding chain e-mails can create problems for e-mail servers. Find out how many total e-mails have been forwarded after 6 levels.

   A 1,111,110   C 1,000,000

   B 6,000,000   D 100,000,000

Forwarded E-mails

Level	E-mails forwarded
1	10
2	100
3	1000
4	10,000

Holt Mathematics

Name _____ Date _____ Class _____

LESSON 4-2 Problem Solving
Look for a Pattern in Integer Exponents

Write the correct answer.

1. The weight of 10^7 dust particles is 1 gram. Evaluate 10^7.

   _____

2. The weight of one dust particle is 10^{-7} gram. Evaluate 10^{-7}.

   _____

3. As of 2001, only 10^6 rural homes in the United States had broadband Internet access. Evaluate 10^6.

   _____

4. Atomic clocks measure time in microseconds. A microsecond is 10^{-6} second. Evaluate 10^{-6}.

   _____

Choose the letter for the best answer.

5. The diameter of the nucleus of an atom is about 10^{-15} meter. Evaluate 10^{-15}.
   A 0.0000000000001
   B 0.00000000000001
   C 0.0000000000000001
   D 0.000000000000001

6. The diameter of the nucleus of an atom is 0.000001 nanometer. How many nanometers is the diameter of the nucleus of an atom?
   F $(-10)^5$
   G $(-10)^6$
   H 10^{-6}
   J 10^{-5}

7. A ruby-throated hummingbird weighs about 3^{-2} ounce. Evaluate 3^{-2}.
   A -9
   B -6
   C $\frac{1}{9}$
   D $\frac{1}{6}$

8. A ruby-throated hummingbird breathes 2×5^3 times per minute while at rest. Evaluate this amount.
   F 1,000
   G 250
   H 125
   J 30

Holt Mathematics

LESSON 4-3
Problem Solving
Properties of Exponents

Write each answer as a power.

1. Cindy separated her fruit flies into equal groups. She estimates that there are 2^{10} fruit flies in each of 2^2 jars. How many fruit flies does Cindy have in all?

2. Suppose a researcher tests a new method of pasteurization on a strain of bacteria in his laboratory. If the bacteria are killed at a rate of 8^9 per sec, how many bacteria would be killed after 8^2 sec?

3. A satellite orbits the earth at about 13^4 km per hour. How long would it take to complete 24 orbits, which is a distance of about 13^5 km?

4. The side of a cube is 3^4 centimeters long. What is the volume of the cube? (Hint: $V = s^3$.)

Use the table to answer Exercises 5–6. The table describes the number of people involved at each level of a pyramid scheme. In a pyramid scheme each individual recruits so many others to participate who in turn recruit others, and so on. Choose the letter of the best answer.

5. Using exponents, how many people will be involved at level 6?

   A 6^6 C 5^5

   B 6^5 D 5^6

Pyramid Scheme

Each person recruits 5 others.

Level	Total Number of People
1	5
2	5^2
3	5^3
4	5^4

6. How many more people will be involved at level 6 than at level 2?

   F 5^4 H 5^5

   G 5^3 J 5^6

7. There are 10^3 ways to make a 3-digit combination, but there are 10^6 ways to make a 6-digit combination. How many times more ways are there to make a 6-digit combination than a 3-digit combination?

   A 5^{10} C 2^5

   B 2^{10} D 10^3

8. After 3 hours, a bacteria colony has $(25^3)^3$ bacteria present. How many bacteria are in the colony?

   F 25^1 H 25^9

   G 25^6 J 25^{33}

Holt Mathematics

Name _____ Date _____ Class _____

Problem Solving
Scientific Notation

Write the correct answer.

1. In June 2001, the Intel Corporation announced that they could produce a silicon transistor that could switch on and off 1.5 trillion times a second. Express the speed of the transistor in scientific notation.

2. With this transistor, computers will be able to do 1×10^9 calculations in the time it takes to blink your eye. Express the number of calculations using standard notation.

3. The elements in this fast transistor are 20 nanometers long. A nanometer is one-billionth of a meter. Express the length of an element in the transistor in meters using scientific notation.

4. The length of the elements in the transistor can also be compared to the width of a human hair. The length of an element is 2×10^{-5} times smaller than the width of a human hair. Express 2×10^{-5} in standard notation.

Use the table to answer Exercises 5–9. Choose the best answer.

5. Express a light-year in miles using scientific notation.
   A 58.8×10^{11} C 588×10^{10}
   B 5.88×10^{12} D 5.88×10^{-13}

6. How many miles is it from Earth to the star Sirius?
   F 4.705×10^{12} H 7.35×10^{12}
   G 4.704×10^{13} J 7.35×10^{11}

7. How many miles is it from Earth to the star Canopus?
   A 3.822×10^{15} C 3.822×10^{14}
   B 1.230×10^{15} D 1.230×10^{14}

Distance From Earth To Stars
Light-Year = 5,880,000,000,000 mi.

Star	Constellation	Distance (light-years)
Sirius	Canis Major	8
Canopus	Carina	650
Alpha Centauri	Centaurus	4
Vega	Lyra	23

8. How many miles is it from Earth to the star Alpha Centauri?
   F 2.352×10^{13} H 2.352×10^{14}
   G 5.92×10^{13} J 5.92×10^{14}

9. How many miles is it from Earth to the star Vega?
   A 6.11×10^{13} C 6.11×10^{14}
   B 1.3524×10^{13} D 1.3524×10^{14}

Holt Mathematics

Problem Solving
LESSON 4-5
Squares and Square Roots

Write the correct answer.

1. For college wrestling competitions, the NCAA requires that the wrestling mat be a square with an area of 1764 square feet. What is the length of each side of the wrestling mat?

2. For high school wrestling competitions, the wrestling mat must be a square with an area of 1444 square feet. What is the length of each side of the wrestling mat?

3. The Japanese art of origami requires folding square pieces of paper. Elena begins with a large sheet of square paper that is 169 square inches. How many squares can she cut out of the paper that are 4 inches on each side?

4. When the James family moved into a new house they had a square area rug that was 132 square feet. In their new house, there are three bedrooms. Bedroom one is 11 feet by 11 feet. Bedroom two is 10 feet by 12 feet and bedroom three is 13 feet by 13 feet. In which bedroom will the rug fit?

Choose the letter for the best answer.

5. A square picture frame measures 36 inches on each side. The actual wood trim is 2 inches wide. The photograph in the frame is surrounded by a bronze mat that measures 5 inches. What is the maximum area of the photograph?

    A 841 sq. inches   **B** 900 sq. inches
    C 1156 sq. inches   **D** 484 sq. inches

6. To create a square patchwork quilt wall hanging, square pieces of material are sewn together to form a larger square. Which number of smaller squares can be used to create a square patchwork quilt wall hanging?

    F 35 squares   **G** 64 squares
    H 84 squares   **J** 125 squares

7. A can of paint claims that one can will cover 400 square feet. If you painted a square with the can of paint, how long would it be on each side?

    A 200 feet   **B** 65 feet
    C 25 feet   **D** 20 feet

8. A box of tile contains 12 tiles. If you tile a square area using whole tiles, how many tiles will you have left from the box?

    F 9   **G** 6
    H 3   **J** 0

Holt Mathematics

Name _____ Date _____ Class _____

Problem Solving
LESSON 4-6 Estimating Square Roots

The distance to the horizon can be found using the formula
$d = 112.88\sqrt{h}$ where d is the distance in kilometers and h is the
number of kilometers from the ground. Round your answer to
the nearest kilometer.

1. How far is it to the horizon when you are standing on the top of Mt. Everest, a height of 8.85 km?

   _____

2. Find the distance to the horizon from the top of Mt. McKinley, Alaska, a height of 6.194 km.

   _____

3. How far is it to the horizon if you are standing on the ground and your eyes are 2 m above the ground?

   _____

4. Mauna Kea is an extinct volcano on Hawaii that is about 4 km tall. You should be able to see the top of Mauna Kea when you are how far away?

   _____

You can find the approximate speed of a vehicle that leaves
skid marks before it stops. The formulas $S = 5.5\sqrt{0.7L}$ and
$S = 5.5\sqrt{0.8L}$, where S is the speed in miles per hour and L is
the length of the skid marks in feet, will give the minimum and
maximum speeds that the vehicle was traveling before the
brakes were applied. Round to the nearest mile per hour.

5. A vehicle leaves a skid mark of 40 feet before stopping. What was the approximate speed of the vehicle before it stopped?

   A 25–35 mph   C 29–31 mph
   B 28–32 mph   D 68–70 mph

6. A vehicle leaves a skid mark of 100 feet before stopping. What was the approximate speed of the vehicle before it stopped?

   F 46–49 mph   H 62–64 mph
   G 50–55 mph   J 70–73 mph

7. A vehicle leaves a skid mark of 150 feet before stopping. What was the approximate speed of the vehicle before it stopped?

   A 50–55 mph   C 55–70 mph
   B 53–58 mph   D 56–60 mph

8. A vehicle leaves a skid mark of 200 feet before stopping. What was the approximate speed of the vehicle before it stopped?

   F 60–63 mph   G 65–70 mph
   H 72–78 mph   J 80–90 mph

Holt Mathematics

LESSON	**Problem Solving**
4-7	*The Real Numbers*

Write the correct answer.

1. Twin primes are prime numbers that differ by 2. Find an irrational number between twin primes 5 and 7.

_____

2. Rounded to the nearest ten-thousandth, $\pi = 3.1416$. Find a rational number between 3 and π.

3. One famous irrational number is *e*. Rounded to the nearest ten-thousandth $e \approx 2.7183$. Find a rational number that is between 2 and *e*.

4. Perfect numbers are those that the divisors of the number sum to the number itself. The number 6 is a perfect number because $1 + 2 + 3 = 6$. The number 28 is also a perfect number. Find an irrational number between 6 and 28.

_____

Choose the letter for the best answer.

5. Which is a rational number?

  A the length of a side of a square with area 2 cm^2

  B the length of a side of a square with area 4 cm^2

  C a non-terminating decimal

  D the square root of a prime number

6. Which is an irrational number?

  F a number that can be expressed as a fraction

  G the length of a side of a square with area 4 cm^2

  H the length of a side of a square with area 2 cm^2

  J the square root of a negative number

7. Which is an integer?

  A the number half-way between 6 and 7

  B the average rainfall for the week if it rained 0.5 in., 2.3 in., 0 in., 0 in., 0 in., 0.2 in., 0.75 in. during the week

  C the money in an account if the balance was $213.00 and $21.87 was deposited

  D the net yardage after plays that resulted in a 15 yard loss, 10 yard gain, 6 yard gain and 5 yard loss

8. Which is a whole number?

  F the number half-way between 6 and 7

  G the total amount of sugar in a recipe that calls for $\frac{1}{4}$ cup of brown sugar and $\frac{3}{4}$ cup of granulated sugar

  H the money in an account if the balance was $213.00 and $21.87 was deposited

  J the net yardage after plays that resulted in a 15 yard loss, 10 yard gain, 6 yard gain and 5 yard loss

Holt Mathematics

Name _____ Date _____ Class _____

Problem Solving
The Pythagorean Theorem

Write the correct answer. Round to the nearest tenth.

1. A utility pole 10 m high is supported by two guy wires. Each guy wire is anchored 3 m from the base of the pole. How many meters of wire are needed for the guy wires?

   _____

2. A 12 foot-ladder is resting against a wall. The base of the ladder is 2.5 feet from the base of the wall. How high up the wall will the ladder reach?

   _____

3. The base-path of a baseball diamond form a square. If it is 90 ft from home to first, how far does the catcher have to throw to catch someone stealing second base?

   _____

4. A football field is 100 yards with 10 yards at each end for the end zones. The field is 45 yards wide. Find the length of the diagonal of the entire field, including the end zones.

   _____

Choose the letter for the best answer.

5. The frame of a kite is made from two strips of wood, one 27 inches long, and one 18 inches long. What is the perimeter of the kite? Round to the nearest tenth.

   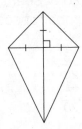

   A 18.8 in.   **C** 65.7 in.
   B 32.8 in.   **D** 131.2 in.

6. The glass for a picture window is 8 feet wide. The door it must pass through is 3 feet wide. How tall must the door be for the glass to pass through the door? Round to the nearest tenth.

   F 3.3 ft   **H** 7.4 ft
   G 6.7ft   **J** 8.5 ft

7. A television screen measures approximately 15.5 in. high and 19.5 in. wide. A television is advertised by giving the approximate length of the diagonal of its screen. How should this television be advertised?

   A 25 in.   **C** 12 in.
   B 21 in.   **D** 6 in.

8. To meet federal guidelines, a wheelchair ramp that is constructed to rise 1 foot off the ground must extend 12 feet along the ground. How long will the ramp be? Round to the nearest tenth.

   F 11.9 ft   **H** 13.2 ft
   G 12.0 ft   **J** 15.0 ft

Holt Mathematics

Problem Solving

LESSON 5-1

Ratios and Proportions

A medicine for dogs indicates that the medicine should be administered in the ratio 0.5 tsp per 5 lb, based on the weight of the dog. Write the correct answer.

1. Jaime has a 60 lb dog. She plans to give the dog 12 teaspoons of medicine. Is she administering the medicine correctly?

2. Jaime also has a 15 lb puppy. She plans to give the puppy 1.5 teaspoons of medicine. Is she administering the medicine correctly?

_____

Sports statistics can be given as ratios. Find the ratios for the given statistics. Reduce each ratio.

3. In 69 games, Darrel Armstrong of the Orlando Magic had 136 steals and 144 turnovers. What is his steals per turnover ratio?

4. In 69 games, Ben Wallace of the Detroit Pistons blocked 234 shots. What is his blocks per game ratio?

_____

Choose the letter for the best answer.

5. There are 675 students and 30 teachers in the middle school. What is the ratio of teachers to students?

   A $\frac{45}{2}$ C $\frac{1}{27}$

   B $\frac{2}{45}$ D $\frac{27}{1}$

6. In a science experiment, out of a sample of seeds, 13 sprouted and 7 didn't. What is the ratio of seeds that sprouted to the number of seeds planted?

   F $\frac{13}{7}$ H $\frac{13}{20}$

   G $\frac{7}{13}$ J $\frac{7}{20}$

7. Many Internet services advertise their customer to modem ratio. One company advertises a 10 to 1 customer to modem ratio. Find a ratio that is equivalent to $\frac{10}{1}$.

   A $\frac{40}{4}$ C $\frac{400}{4}$

   B $\frac{2}{20}$ D $\frac{50}{10}$

8. A molecule of sulfuric acid contains 2 atoms of hydrogen to every 4 atoms of oxygen. Which combination of hydrogen and oxygen atoms could be sulfuric acid?

   F 4 atoms of hydrogen and 6 atoms of oxygen

   G 6 atoms of hydrogen and 10 atoms of oxygen

   H 6 atoms of hydrogen and 12 atoms of oxygen

   J 16 atoms of hydrogen and 8 atoms of oxygen

Holt Mathematics

Problem Solving

LESSON 5-2

Ratios, Rates, and Unit Rates

Scientists have researched the ratio of brain weight to body size in different animals. The results are in the table below.

1. Order the animals by their brain weight to body weight ratio, from smallest to largest.

   _____

   _____

2. It has been hypothesized that the higher the brain weight to body weight ratio, the more intelligent the animal is. By this measure, which animals listed are the most intelligent?

   _____

3. Name two sets of animals that have approximately the same brain weight to body weight ratio.

   _____

Animal	Brain Weight / Body Weight
Cat	$\frac{1}{100}$
Dog	$\frac{1}{125}$
Elephant	$\frac{1}{560}$
Hippo	$\frac{1}{2789}$
Horse	$\frac{1}{600}$
Human	$\frac{1}{40}$
Small birds	$\frac{1}{12}$

Find the unit rate. Round to the nearest hundredth.

4. A 64-ounce bottle of apple juice costs $1.35.

   A $0.01/oz   **C** $0.47/oz

   B $0.02/oz   **D** $47.4/oz

5. Find the unit rate for a 2 lb package of hamburger that costs $3.45.

   F $0.58/lb   **H** $1.73/lb

   G $1.25/b   **J** $2.28/b

6. 12 slices of pizza cost $9.00.

   A $0.45/slice   **C** $0.75/slice

   B $0.50/slice   **D** $1.33/slice

7. John is selling 5 comic books for $6.00.

   F $0.83/book   **H** $1.02/book

   G $1.20/book   **J** $1.45/book

8. There are 64 beats in 4 measures of music.

   A 16 beats/measure

   B 12 beats/measure

   C 4 beats/measure

   D 0.06 beats/measure

9. The average price of a 30 second commercial for the 2002 Super Bowl was $1,900,000.

   F $120.82/sec

   G $1,242.50/sec

   H $5,839.02/sec

   J $63,333.33/sec

Holt Mathematics

Problem Solving
5-3 *Dimensional Analysis*

Use the following: 1 mile = 1.609 km; 1 kg = 2.2046 lb. Round to the nearest tenth.

1. Worker bees travel up to 14 km to find pollen and nectar. How far will a worker bee travel in miles?

2. Worker bees can travel at 24 km/h. How fast can the worker bee travel in miles per hour?

_____

3. The average hippopotamus weighs 1800 kg. How many pounds does the average hippopotamus weigh?

4. At the age of 45, an elephant grows teeth, each weighing 4 kg. How many pounds do these teeth weigh?

_____

Paraceratherium was the biggest land mammal there has ever been. It lived about 35 million years ago and was 8 m tall and 11 m long. It looked like a gigantic rhinoceros but had a long neck like a giraffe. 1 foot = 0.3048 meters. Round to the nearest tenth.

5. How tall was the paraceratherium in feet?

6. How long was the paraceratherium in feet?

_____

Round to the nearest tenth. Choose the letter for the best answer.

7. The fastest sporting animal is the racing pigeon that flies up to 110 mi an hour. How fast is the racing pigeon in feet each second?

   A 75.0 ft/s   C 543.2 ft/s
   B 161.3 ft/s   D 9,680 ft/s

8. The longest gloved fight between two Americans lasted for more than seven hours before being declared a draw. How many seconds did the fight last?

   F 127 s   H 420 s
   G 385 s   J 25,200 s

9. The average person falls asleep in seven minutes. How many seconds does it take the average person to fall asleep?

   A 127 s   C 420 s
   B 385 s   D 25,200 s

10. The brain of an average adult male weighs 55 oz. How many pounds does the average male brain weigh?

   F 3.4 lb   H 13.8 lb
   G 5.8 lb   J 880 lb

Holt Mathematics

LESSON
5-4

Problem Solving
Solving Proportions

Use the ratios in the table to answer each question. Round to the nearest tenth.

Body Part	$\dfrac{\text{Body Part}}{\text{Height}}$
Femur	$\dfrac{1}{4}$
Tibia	$\dfrac{1}{5}$
Hand span	$\dfrac{2}{17}$
Arm span	$\dfrac{1}{1}$
Head circumference	$\dfrac{1}{3}$

1. Which body part is the same length as the person's height?

   _____

2. If a person's tibia is 13 inches, how tall would you expect the person to be?

   _____

3. If a person's hand span is 8.5 inches, about how tall would you expect the person to be?

   _____

4. If a femur is 18 inches long, how many feet tall would you expect the person to be?

   _____

5. What would you expect the head circumference to be of a person who is 5.5 feet tall?

   _____

6. What would you expect the hand span to be of a person who is 5 feet tall?

   _____

Choose the letter for the best answer.

7. Five milliliters of a children's medicine contains 400 mg of the drug amoxicillin. How many mg of amoxicillin does 25 mL contain?

   A 0.3 mg   **C** 2000 mg

   B 80 mg   **D** 2500 mg

8. Vladimir Radmanovic of the Seattle Supersonics makes, on average, about 2 three-pointers for every 5 he shoots. If he attempts 10 three-pointers in a game, how many would you expect him to make?

   F 4   **H** 8

   G 5   **J** 25

9. In 2002, a 30-second commercial during the Super Bowl cost an average of $1,900,000. At this rate, how much would a 45-second commercial cost?

   A $1,266,666   **C** $3,500,000

   B $2,850,000   **D** $4,000,000

10. A medicine for dogs indicates that the medicine should be administered in the ratio 2 teaspoons per 5 lb, based on the weight of the dog. How much should be given to a 70 lb dog?

   F 5 teaspoons   **H** 14 teaspoons

   G 12 teaspoons   **J** 28 teaspoons

Holt Mathematics

Name _____ Date _____ Class _____

Problem Solving
Similar Figures

Write the correct answer.

1. Until 1929, United States currency measured 3.13 in. by 7.42 in. The current size is 2.61 in. by 6.14 in. Are the bills similar?

_____

2. Owen has a 3 in. by 5 in. photograph. He wants to make it as large as he can to fit in a 10 in. by 12.5 in. ad. What scale factor will he use? What will be the new size?

_____

3. A painting is 15 cm long and 8 cm wide. In a reproduction that is similar to the original painting, the length is 36 cm. How wide is the reproduction?

_____

4. The two shortest sides of a right triangle are 10 in. and 24 in. long. What is the length of the shortest side of a similar right triangle whose two longest sides are 36 in. and 39 in.?

_____

The scale on a map is 1 inch = 40 miles. Round to the nearest mile.

5. On the map, it is 5.75 inches from Orlando to Miami. How many miles is it from Orlando to Miami?

   A 46 miles   **C** 230 miles

   B 175 miles   **D** 340 miles

6. On the map it is $18\frac{1}{8}$ inches from Norfolk, VA, to Indianapolis, IN. How many miles is it from Norfolk to Indianapolis?

   F 58 miles   **H** 800 miles

   G 725 miles   **J** 1025 miles

7. It is 185 miles from Chicago to Indianapolis. On the map it is 2.5 inches from Indianapolis to Terra Haute, IN. How far is it from Chicago to Terra Haute going through Indianapolis?

   A 100 miles   **C** 430 miles

   B 285 miles   **D** 7500 miles

8. On the map, it is 7.5 inches from Chicago to Cincinnati. Traveling at 65 mi/h, how long will it take to drive from Chicago to Cincinnati? Round to the nearest tenth of an hour.

   F 4.6 hours   **H** 8.7 hours

   G 5.2 hours   **J** 12.0 hours

Holt Mathematics

Name _____ Date _____ Class _____

Problem Solving
Dilations

Write the correct answer.

1. When you enlarge something on a photocopy machine, is the image a dilation?

   _____

2. When you make a photocopy that is the same size, is the image a dilation? If so, what is the scale factor?

   _____

3. In the movie *Honey, I Blew Up the Kid*, a two-year-old-boy is enlarged to a height of 112 feet. If the average height of a two-year old boy is 3 feet, what is the scale factor of this enlargement?

   _____

4. In the movie *Honey, I Shrunk the Kids*, an inventor shrinks his kids by a scale factor of about $\frac{1}{240}$. If his kids were about 5 feet tall, how many inches tall were they after they were shrunk?

   _____

Use the coordinate plane for Exercises 5–6. Round to the nearest tenth. Choose the letter for the best answer.

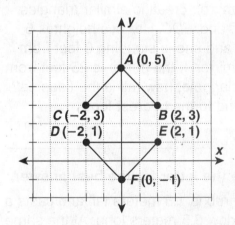

5. What will be the coordinates of *A'*, *B'* and *C'* after △*ABC* is dilated by a factor of 5?

   A *A'*(5, 10), *B'*(7, 8), *C'*(3, 8)

   B *A'*(0, 25), *B'*(10, 15), *C'*(−10, 15)

   C *A'*(0, 5), *B'*(10, 3), *C'*(−10, 3)

   D *A'*(0, 15), *B'*(2, 15), *C'*(−2, 15)

6. What will be the coordinates of *D'*, *E'* and *F'* after △*DEF* is dilated by a factor of 5?

   F *D'*(−10, 5), *E'*(10, 5), *F'*(0, −5)

   G *D'*(10, 5), *E'*(5, 5), *F'*(0, 5)

   H *D'*(10, 0), *E'*(5, 0), *F'*(−5, 0)

   J *D'*(10, 5), *E'*(5, 5), *F'*(−10, 0)

7. The projection of a movie onto a screen is a dilation. The universally accepted film size for movies has a width of 35 mm. If a movie screen is 12 m wide, what is the dilation factor?

   A 420   **C** 342.9

   B 0.3   **D** 2916.7

Holt Mathematics

Name _____ Date _____ Class _____

Problem Solving
Indirect Measurement

Write the correct answer.

1. Celine wants to know the width of the pond. She drew the diagram shown below and labeled it with the measurements she made. How wide is the pond?

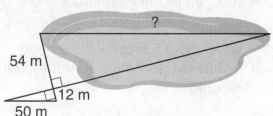

2. Vince wants to know the distance across the canyon. He drew the diagram and labeled it with the measurements he made. What is the distance across the canyon?

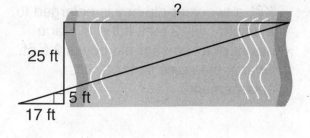

3. Paula places a mirror between herself and a flagpole. She stands so she can see the top of the flagpole in the mirror, creating similar triangles *ABC* and *EDC*. Her eye height is 5 feet and she is standing 6 feet from the mirror. If the mirror is 25 feet from the flagpole, how tall is the flagpole? Round to the nearest foot.

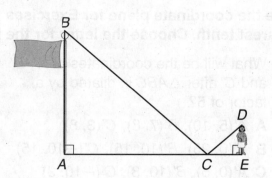

Choose the letter for the best answer.

4. A shrub is 1.5 meters tall and casts a shadow 3.5 meters long. At the same time, a radio tower casts a shadow 98 meters long. How tall is the radio tower?

   A 33 m

   B 42 m

   C 147 m

   D 329 m

5. Kim is 56 inches tall. His friend is 42 inches tall. Kim's shadow is 24 inches long. How long is his friend's shadow at the same time?

   F 18 in.

   G 32 in.

   H 38 in.

   J 98 in.

Holt Mathematics

LESSON 5-8
Problem Solving
Scale Drawings and Scale Models

Round to the nearest tenth. Write the correct answer.

1. The Statue of Liberty is approximately 305 feet tall. A scale model of the Statue of Liberty is 5 inches tall. The scale of the model is 1 in. : _____ ft.

   _____

2. The right arm of the Statue of Liberty is 42 feet long. How long is the right arm of the Statue of Liberty model given in Exercise 1?

3. The diameter of an atom is 10^{-9} cm. If a scale drawing of an atom has a diameter of 10 cm, the scale of the drawing is 1 cm : _____ cm.

   _____

4. The diameter of the nucleus of an atom is 10^{-13} cm. If a scale drawing of the nucleus of an atom has a diameter of 1 cm, the scale of the drawing is 1 cm : _____ cm.

   _____

The toy car in Exercises 5-6 has a scale of $\frac{1}{40}$. Choose the letter for the best answer.

5. The diameter of the steering wheel of the actual car is 15 inches. What is the diameter of the toy car's steering wheel?

   A $\frac{3}{8}$ in.   C $1\frac{1}{2}$ in.

   B $\frac{1}{2}$ in.   D $2\frac{2}{3}$ in.

6. The diameter of the toy car's tire is $\frac{5}{8}$ in. What is the diameter of the tire of the actual car?

   F $12\frac{1}{2}$ in.   H 25 in.

   G 16 in.   J 64 in.

7. On a scale drawing, a 14 ft room is pictured as 3.5 inches. What is the scale of the drawing?

   A $\frac{1}{4}$ in.   C $\frac{1}{4}$:14

   B $\frac{1}{2}$ in.   D 1:56

8. On a scale drawing, $\frac{1}{2}$ inch = 1 foot. A room is pictured as 7.5 inches by 6 inches. How many square yards of carpet are needed for the room?

   F 5 yd^2 H 45 yd^2

   G 20 yd^2 J 90 yd^2

9. On a scale drawing of a computer component, $\frac{1}{4}$ in. = 4 in. On the drawing, a piece is $\frac{3}{8}$ in. long. How long is the actual piece?

   A 1.5 in.   C 6 in.

   B 3 in.   D 7.5 in.

10. A scale drawing has a $\frac{1}{4}$ inch scale. The width of a 12 foot room is going to be increased by 4 feet. How much wider will the room be on the drawing?

   F $\frac{1}{4}$ in.   H 1 in.

   G $\frac{1}{2}$ in.   J 4 in.

Holt Mathematics

LESSON 6-1

Problem Solving
Relating Decimals, Fractions, and Percents

The table shows the ratio of brain weight to body size in different animals. Use the table for Exercises 1–3.
Write the correct answer.

1. Complete the table to show the percent of each animal's body weight that is brain weight. Round to the nearest hundredth.

2. Which animal has a greater brain weight to body size ratio, a dog or an elephant?

_____

3. List the animals from least to greatest brain weight to body size ratio.

_____

Animal	$\dfrac{\text{Brain Weight}}{\text{Body Weight}}$	Percent
Mouse	$\dfrac{1}{40}$	
Cat	$\dfrac{1}{100}$	
Dog	$\dfrac{1}{125}$	
Horse	$\dfrac{1}{600}$	
Elephant	$\dfrac{1}{560}$	

The table shows the number of wins and losses of the top teams in the National Football Conference from 2004. Choose the letter of the best answer. Round to the nearest tenth.

4. What percent of games did the Green Bay Packers win?

   A 10%   C 37.5%

   B 60%   D 62.5%

Team	Wins	Losses
Philadelphia Eagles	13	3
Green Bay Packers	10	6
Atlanta Falcons	11	5
Seattle Seahawks	9	7

5. Which decimal is equivalent to the percent of games the Seattle Seahawks won?

   F 0.05625   H 5.625

   G 0.5625   J 56.25

6. Which team listed had the highest percentage of wins?

   A Philadelphia Eagles

   B Green Bay Packers

   C Atlanta Falcons

   D Seattle Seahawks

Holt Mathematics

LESSON 6-2 **Problem Solving**
Estimate with Percents

Write an estimate.

1. A store requires you to pay 15% up front on special orders. If you plan to special order items worth $74.86, estimate how much you will have to pay up front.

_____

2. A store is offering 25% off of everything in the store. Estimate how much you will save on a jacket that is normally $58.99.

_____

3. A certain kind of investment requires that you pay a 10% penalty on any money you remove from the investment in the first 7 years. If you take $228 out of the investment, estimate how much of a penalty you will have to pay.

_____

4. John notices that about 18% of the earnings from his job go to taxes. If he works 14 hours at $6.25 an hour, about how much of his check will go for taxes?

_____

Choose the letter for the best estimate.

5. In its first week, an infant can lose up to 10% of its body weight in the first few days of life. Which is a good estimate of how many ounces a 5 lb 13 oz baby might lose in the first week of life?

   A 0.6 oz   C 18 oz

   B 9 oz   D 22 oz

6. A CD on sale costs $12.89. Sales tax is 4.75%. Which is the best estimate of the total cost of the CD?

   F $13.30   H $14.20

   G $13.55   J $14.50

7. In a class election, Pedro received 52% of the votes. There were 274 students who voted in the election. Which is the best estimate of the number of students who voted for Pedro?

   A 70 students   C 125 students

   B 100 students   D 140 students

8. Mel's family went out for breakfast. The bill was $24.25 plus 5.2% sales tax. Mel wants to leave a 20% tip. Which is the best estimate of the total bill?

   F $25.45   H $30.25

   G $29.25   J $32.25

Holt Mathematics

Name _____ Date _____ Class _____

Problem Solving
Finding Percents

Write the correct answer.

1. Florida State University in Tallahassee, Florida has 29,820 students. Approximately 60% of the students are women. How many of the students are women?

2. The yearly cost of tuition, room and board at Florida State University for a Florida resident is $10,064. If tuition is $3,208 a year, what percent of the yearly cost is tuition? Round to the nearest tenth of a percent.

3. The yearly cost of tuition, room and board at Florida State University for a non-Florida resident is $23,196. If tuition is $16,340 a year, what percent of the yearly cost is tuition for a non-resident? Round to the nearest tenth of a percent.

4. Approximately 65% of the students who apply to Florida State University are accepted. If 15,000 students apply to Florida State University, how many would you expect to be accepted?

The top four NBA field goal shooters for the 2003–2004 regular season are given in the table below. Choose the letter for the best answer.

5. What percent of field goals did Shaquille O'Neal make? Round to the nearest tenth of a percent.
   - **A** 0.6%
   - **B** 1.71%
   - **C** 58.4%
   - **D** 59.2%

**NBA Field Goal Leaders
2003–2004 Season**

Player	Attempts	Made	Percent
Shaquille O'Neal	948	554	
Donyell Marshall	604		56.6
Elton Brand	950	348	
Dale Davis	913		53.5

6. How many field goals did Donyell Marshall make in the 2003–2004 regular season?
   - **F** 38
   - **G** 114
   - **H** 295
   - **J** 342

7. What percent of field goals did Elton Brand make? Round to the nearest tenth of a percent.
   - **A** 1.87%
   - **B** 50%
   - **C** 51.9%
   - **D** 36.6%

8. How many field goals did Dale Davis make in the 2003–2004 regular season?
   - **F** 274
   - **G** 378
   - **H** 457
   - **J** 488

Holt Mathematics

Problem Solving

LESSON 6-4

Finding a Number When the Percent is Known

Write the correct answer.

1. The two longest running Broadway shows are *Cats* and *A Chorus Line*. *A Chorus Line* had 6137, or about 82% of the number of performances that *Cats* had. How many performances of *Cats* were there?

2. *Titanic* and *Star Wars* have made the most money at the box office. *Star Wars* made about 76.7% of the money that *Titanic* made at the box office. If *Star Wars* made about $461 million, how much did *Titanic* make? Round to the nearest million dollars.

Use the table below. Round to the nearest tenth of a percent.

3. What percent of students are in Pre-K through 8th grade?

_____

4. What percent of students are in grades 9–12?

_____

Public Elementary and Secondary School Enrollment, 2001

Grades	Population (in thousands)
Pre-K through grade 8	33,952
Grades 9–12	13,736
Total	47,688

Choose the letter for the best answer.

5. In 2000, women earned about 72.2% of what men did. If the average woman's weekly earnings was $491 in 2000, what was the average man's weekly earnings? Round to the nearest dollar.

   A $355   C $680

   B $542   D $725

6. The highest elevation in North America is Mt. McKinley at 20,320 ft. The highest elevation in Australia is Mt. Kosciusko, which is about 36% of the height of Mt. McKinley. What is the highest elevation in Australia? Round to the nearest foot.

   F 5480 ft   H 12,825 ft

   G 7315 ft   J 56,444 ft

7. The Gulf of Mexico has an average depth of 4,874 ft. This is about 36.2% of the average depth of the Pacific Ocean. What is the average depth of the Pacific Ocean? Round to the nearest foot.

   A 1764 ft   C 10,280 ft

   B 5843 ft   D 13,464 ft

8. Karl Malone is the NBA lifetime leader in free throws. He attempted 11,703 and made 8,636. What percent did he make? Round to the nearest tenth of a percent.

   F 1.4%   H 73.8%

   G 58.6%   J 135.6%

Holt Mathematics

Name _____ Date _____ Class _____

Problem Solving
Percent Increase and Decrease

Use the table below. Write the correct answer.

1. What is the percent increase in the population of Las Vegas, NV from 1990 to 2000? Round to the nearest tenth of a percent.

_____

2. What is the percent increase in the population of Naples, FL from 1990 to 2000? Round to the nearest tenth of a percent.

_____

Fastest Growing Metropolitan Areas, 1990–2000

Metropolitan Area	Population		Percent Increase
	1990	2000	
Las Vegas, NV	852,737	1,563,282	
Naples, FL	152,099	251,377	
Yuma, AZ	106,895		49.7%
McAllen-Edinburg-Mission, TX	383,545		48.5%

3. What was the 2000 population of Yuma, AZ to the nearest whole number?

_____

4. What was the 2000 population of McAllen-Edinburg-Mission, TX metropolitan area to the nearest whole number?

_____

For exercises 5–7, round to the nearest tenth. Choose the letter for the best answer.

5. The amount of money spent on automotive advertising in 2000 was 4.4% lower than in 1999. If the 1999 spending was $1812.3 million, what was the 2000 spending?

A $79.7 million **C** $1892 million

B $1732.6 million **D** $1923.5 million

6. In 1967, a 30-second Super Bowl commercial cost $42,000. In 2000, a 30-second commercial cost $1,900,000. What was the percent increase in the cost?

F 1.7% **H** 442.4%

G 44.2% **J** 4423.8%

7. In 1896 Thomas Burke of the U.S. won the 100-meter dash at the Summer Olympics with a time of 12.00 seconds. In 2004, Justin Gatlin of the U.S. won with a time of 9.85 seconds. What was the percent decrease in the winning time?

A 2.15% **C** 21.8%

B 17.9% **D** 45.1%

8. In 1928 Elizabeth Robinson won the 100-meter dash with a time of 12.20 seconds. In 2004, Yuliya Nesterenko won with a time that was about 10.4% less than Robinson's winning time. What was Nesterenko's time, rounded to the nearest hundredth?

F 9.83 seconds **H** 12.16 seconds

G 10.93 seconds **J** 13.47 seconds

Holt Mathematics

Problem Solving
LESSON 6-6 *Applications of Percents*

Write the correct answer.

1. The sales tax rate for a community is 6.75%. If you purchase an item for $500, how much will you pay in sales tax?

2. A community is considering increasing the sales tax rate 0.5% to fund a new sports arena. If the tax rate is raised, how much more will you pay in sales tax on $500?

3. Trent earned $28,500 last year. He paid $8,265 for rent. What percent of his earnings did Trent pay for rent?

4. Julie has been offered two jobs. The first pays $400 per week. The second job pays $175 per week plus 15% commission on her sales. How much will she have to sell in order for the second job to pay as much as the first?

Choose the letter for the best answer. Round to the nearest cent.

5. Clay earned $2,600 last month. He paid $234 for entertainment. What percent of his earnings did Clay pay in entertainment expenses?

    A 9%

    B 11%

    C 30%

    D 90%

6. Susan's parents have offered to help her pay for a new computer. They will pay 30% and Susan will pay 70% of the cost of a new computer. Susan has saved $550 for a new computer. With her parents help, how expensive of a computer can she afford?

    F $165.00   H $1650.00

    G $785.71   J $1833.33

7. Kellen's bill at a restaurant before tax and tip is $22.00. If tax is 5.25% and he wants to leave 15% of the bill including the tax for a tip, how much will he spend in total?

    A $22.17   C $26.63

    B $26.46   D $27.82

8. The 8th grade class is trying to raise money for a field trip. They need to raise $600 and the fundraiser they have chosen will give them 20% of the amount that they sell. How much do they need to sell to raise the money for the field trip?

    F $120.00   H $3000.00

    G $857.14   J $3200.00

Holt Mathematics

Name _____ Date _____ Class _____

Problem Solving
Simple Interest

Write the correct answer.

1. Joanna's parents agree to loan her the money for a car. They will loan her $5,000 for 5 years at 5% simple interest. How much will Joanna pay in interest to her parents?

_____

2. How much money will Joanna have spent in total on her car with the loan described in exercise 1?

3. A bank offers simple interest on a certificate of deposit. Jaime invests $500 and after one year earns $40 in interest. What was the interest rate on the certificate of deposit?

_____

4. How long will Howard have to leave $5000 in the bank to earn $250 in simple interest at 2%?

Jan and Stewart Jones plan to borrow $20,000 for a new car. They are trying to decide whether to take out a 4-year or 5-year simple interest loan. The 4-year loan has an interest rate of 6% and the 5-year loan has an interest rate of 6.25%. Choose the letter for the best answer.

5. How much will they pay in interest on the 4-year loan?
   A $4500   C $5000
   B $4800   D $5200

6. How much will they repay with the 4-year loan?
   F $24,500   H $25,000
   G $24,800   J $25,200

7. How much will they pay in interest on the 5-year loan?
   A $5000   C $6250
   B $6000   D $6500

8. How much will they repay with the 5-year loan?
   F $25,000   H $26,250
   G $26,000   J $26,500

9. How much more interest will they pay with the 5-year loan?
   A $1000
   B $1450
   C $1500
   D $2000

10. If the Stewarts can get a 5-year loan with 5.75% simple interest, which of the loans is the best deal?
    F 4 year, 6%
    G 5 year, 5.75%
    H 5 year, 6.25%
    J Cannot be determined

Holt Mathematics

Name _____ Date _____ Class _____

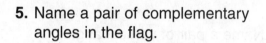

Problem Solving
7-1 Points, Lines, Planes, and Angles

Use the flag of the Bahamas to solve the problems.

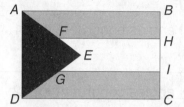

1. Name four points in the flag.

   _____

2. Name four segments in the flag.

   _____

3. Name a right angle in the flag.

   _____

4. Name two acute angles in the flag.

   _____

5. Name a pair of complementary angles in the flag.

   _____

6. Name a pair of supplementary angles in the flag.

   _____

The diagram illustrates a ray of light being reflected off a mirror. The angle of incidence is congruent to the angle of reflection. Choose the letter for the best answer.

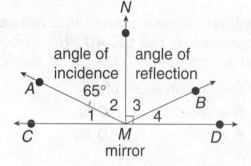

7. Name two rays in the diagram.
   A $\overleftrightarrow{AM}$, $\overleftrightarrow{MB}$ C $\overrightarrow{MA}$, $\overrightarrow{MB}$
   B $\overrightarrow{MA}$, $\overrightarrow{BM}$ D $\overrightarrow{MA}$, $\overleftrightarrow{MB}$

8. Name a pair of complementary angles.
   F ∠NMB, ∠BMD   H ∠CMA, ∠AMD
   G ∠AMN, ∠NMB   J ∠CMA, ∠DMB

9. Which angle is congruent to ∠2?
   A ∠1   C ∠3
   B ∠4   D none

10. Find the measure of ∠4.
    F 65°   H 25°
    G 35°   J 90°

11. Find the measure of ∠1.
    A 65°   C 25°
    B 35°   D 90°

12. Find the measure of ∠3.
    F 90°   H 35°
    G 45°   J 65°

Holt Mathematics

LESSON 7-2 Problem Solving
Parallel and Perpendicular Lines

The figure shows the layout of parking spaces in a parking lot.
$\overline{AB} \parallel \overline{CD} \parallel \overline{EF}$

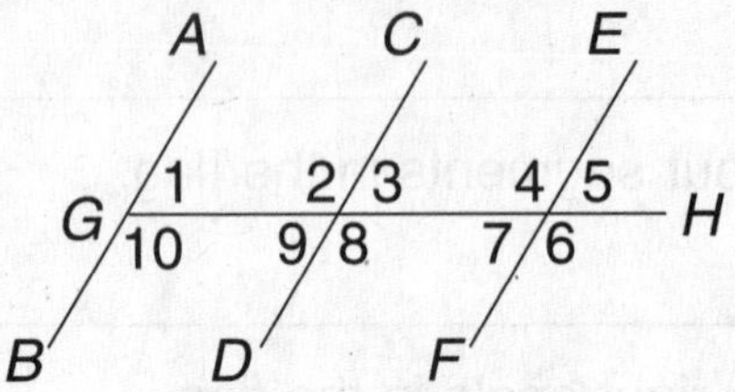

1. Name all angles congruent to ∠1.

_____

2. Name all angles congruent to ∠2.

_____

3. Name a pair of supplementary angles.

_____

4. If m∠1 = 75°, find the measures of the other angles.

_____

_____

5. Name a pair of vertical angles.

_____

6. If m∠1 = 90°, then $\overline{GH}$ is perpendicular to

_____.

The figure shows a board that will be cut along parallel
segments GB and CF. $\overline{AD} \parallel \overline{HE}$. Choose the letter for the
best answer.

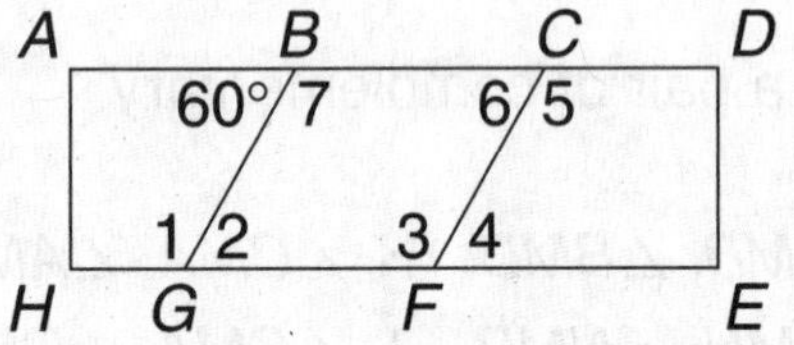

7. Find the measure of ∠1.

   A 45°   **C** 60°

   B 120°   **D** 90°

8. Find the measure of ∠2.

   F 30°   **H** 60°

   G 120°   **J** 90°

9. Find the measure of ∠3.

   A 30°   **C** 60°

   B 120°   **D** 90°

10. Find the measure of ∠4.

   F 45°   **H** 60°

   G 120°   **J** 90°

11. Find the measure of ∠5.

   A 30°   **C** 60°

   B 120°   **D** 90°

12. Find the measure of ∠6.

   F 30°   **H** 60°

   G 120°   **J** 90°

13. Find the measure of ∠7.

   A 45°   **C** 60°

   B 120°   **D** 90°

Holt Mathematics

Name _____ Date _____ Class _____

Problem Solving
Angles in Triangles

The American flag must be folded according to certain rules
that result in the flag being folded into the shape of a triangle.
The figure shows a frame designed to hold an American flag.

1. Is the triangle acute, right, or obtuse?

   _____

2. Is the triangle equilateral, isosceles,
   or scalene?

   _____

3. Find $x°$.   4. Find $y°$.

   _____   _____

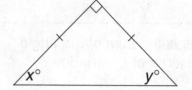

**The figure shows a map of three streets. Choose the letter for
the best answer.**

5. Find $x°$.
   A 22°   **C** 30°
   B 128°   **D** 68°

6. Find $w°$.
   F 22°   **H** 30°
   G 128°   **J** 52°

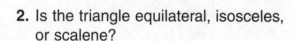

7. Find $y°$.   8. Find $z°$.
   A 22°   **F** 22°
   B 30°   **G** 30°
   C 128°   **H** 128°
   D 143°   **J** 143°

9. Which word best describes the   10. Which word best describes the
   triangle formed by the streets?   triangle formed by the streets?
   A acute   **F** equilateral
   B right   **G** isosceles
   C obtuse   **H** scalene
   D equilateral   **J** acute

                   Holt Mathematics

Name _____ Date _____ Class _____

The figure shows how the glass for a window will be cut from a square piece. Cuts will be made along $\overline{CE}$, $\overline{FH}$, $\overline{IK}$, and $\overline{LB}$.

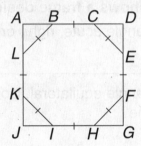

1. What shape is the window?

   _____

2. What is the sum of the angle measures of the window?

   _____

3. What is the measure of each angle of the window?

   _____

4. Based on the angles, what kind of triangle is $\triangle CDE$?

   _____

5. Based on the sides, what kind of triangle is $\triangle CDE$?

   _____

The figure shows how parallel cuts will be made along $\overline{AD}$ and $\overline{BC}$. $\overline{AB}$ and $\overline{CD}$ are parallel. Choose the letter for the best answer.

6. Which word correctly describes figure *ABCD* after the cuts are made?

   A triangle

   B quadrilateral

   C pentagon

   D hexagon

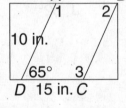

7. Which word correctly describes figure *ABCD* after the cuts are made?

   F parallelogram

   G trapezoid

   H rectangle

   J rhombus

8. Find the measure of $\angle 1$.

   A 45°

   B 65°

   C 90°

   D 115°

9. Find the measure of $\angle 2$.

   F 45°   **H** 65°

   G 90°   **J** 115°

10. Find the measure of $\angle 3$.

   A 45°   **C** 65°

   B 90°   **D** 115°

Holt Mathematics

Name _____ Date _____ Class _____

Problem Solving
Coordinate Geometry

The Uniform Federal Accessibility Standards describes the
standards for making buildings accessible for the handicapped. The
standards say that the least possible slope should be used for a

ramp and that the maximum slope of a ramp should be $\frac{1}{12}$.

1. What is the slope of the pictured
ramp? Does the ramp meet the
standard?

ramp
12 in.
12 ft

2. What is the slope of the pictured
ramp? Does the ramp meet the
standard?

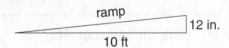

ramp
12 in.
10 ft

_____

_____

Write the correct answer.

3. Find the slope of the roof.

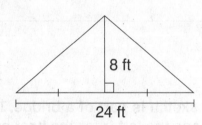

8 ft

24 ft

_____

Choose the letter that represents the slope.

4. Many building codes require that a
staircase be built with a maximum
rise of 8.25 inches for a minimum
tread width (run) of 9 inches.

  A $\frac{8}{9}$ **C** $\frac{9}{8.25}$

  B $\frac{11}{12}$ **D** $\frac{12}{11}$

5. Hills that have a rise of about 10 feet
for every 17 feet horizontally are too
steep for most cars.

  F $\frac{10}{17}$ **H** $\frac{17}{10}$

  G $\frac{2}{5}$ **J** $\frac{3}{5}$

6. At its steepest part, an intermediate
ski run has a rise of about 4 feet for
10 feet horizontally.

  A $\frac{2}{5}$ **C** $\frac{5}{2}$

  B $\frac{4}{5}$ **D** $\frac{5}{4}$

7. Black diamond, or expert, ski slopes
often have a rise of 10 feet for every
14 feet horizontally.

  F $\frac{7}{5}$ **H** $\frac{5}{7}$

  G $\frac{2}{7}$ **J** $\frac{7}{2}$

Holt Mathematics

LESSON
7-6

Problem Solving
Congruence

Use the American patchwork quilt block design called Carnival
to answer the questions. Triangle *AIH* ≅ Triangle *AIB*,
Triangle *ACJ* ≅ Triangle *AGJ*, Triangle *GFJ* ≅ Triangle *CDJ*.

1. What is the measure of ∠*IAB*?

_____

2. What is the measure of $\overline{AH}$?

_____

3. What is the measure of $\overline{AG}$?

_____

4. What is the measure of ∠*JDC*?

_____

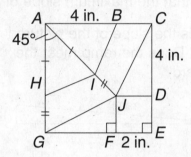

5. What is the measure of $\overline{FG}$?

_____

The sketch is part of a bridge. Trapezoid *ABEF* ≅ Trapezoid *DEBC*.
Choose the letter for the best answer.

6. What is the measure of $\overline{DE}$?

   A 4 feet

   B 8 feet

   C 16 feet

   D Cannot be determined

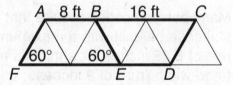

7. What is the measure of $\overline{FE}$?

   F 4 feet   **H** 8 feet

   G 16 feet   **J** 24 feet

8. What is the measure of ∠*FAB*?

   A 45°   **C** 60°

   B 90°   **D** 120°

9. What is the measure of ∠*ABE*?

   F 45°   **H** 60°

   G 90°   **J** 120°

10. What is the measure of ∠*EBC*?

   A 45°   **C** 60°

   B 90°   **D** 120°

11. What is the measure of ∠*BED*?

   F 45°   **H** 60°

   G 90°   **J** 120°

12. What is the measure of ∠*BCD*?

   A 45°   **C** 60°

   B 90°   **D** 120°

Holt Mathematics

Name _____ Date _____ Class _____

Problem Solving
Transformations

Parallelogram ABCD has vertices A(–3, 1), B(–2, 4), C(3, 4), and
D(2, 1). Refer to the parallelogram to write the correct answer.

1. What are the coordinates of point A
   after a reflection across the x-axis?

   _____

2. What are the coordinates of point B
   after a reflection across the y-axis?

   _____

3. What are the coordinates of point C
   after a translation 2 units down?

   _____

4. What are the coordinates of point D
   after a 180° rotation around (0, 0)?

   _____

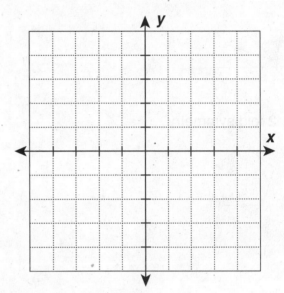

Identify each as a translation, rotation, reflection or none of these.

5.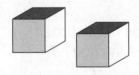

   A translation   **C** rotation
   B reflection   **D** none of these

6. *John ndoↄ*

   F translation   **H** rotation
   G reflection   **J** none of these

7.

   A translation   **C** rotation
   B reflection   **D** none of these

8.

   F translation   **H** rotation
   G reflection   **J** none of these

Holt Mathematics

Name _____ Date _____ Class _____

Problem Solving
Symmetry

Complete the figure. A dashed line is a line of symmetry and a point is a center of rotation.

1.

2.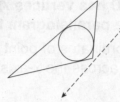

3. 2 fold symmetry

4. 4 fold symmetry

Use the flag of Switzerland to answer the questions.

5. Which of the following would NOT be a line of symmetry?

   A $\overline{HD}$ **C** $\overline{AE}$

   B $\overline{BF}$ **D** $\overline{HB}$

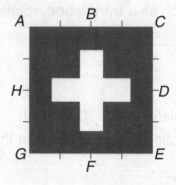

6. How many lines of symmetry does the flag have?

   F 2   **H** 4

   G 6   **J** 8

7. How many folds of rotational symmetry does the flag have?

   A 0   **C** 2

   B 4   **D** 8

8. Which lists all lines of symmetry of the flag?

   F $\overline{AE}$, $\overline{GC}$

   G $\overline{HD}$, $\overline{BF}$

   H $\overline{HD}$, $\overline{BF}$, $\overline{AE}$, $\overline{GC}$

   J $\overline{HB}$, $\overline{DF}$, $\overline{AE}$, $\overline{GC}$

9. Which describes the center of rotation?

   A intersection of $\overline{BF}$ and $\overline{HD}$

   B intersection of $\overline{AE}$ and $\overline{HB}$

   C A

   D There is no center of rotation

Holt Mathematics

Name _____ Date _____ Class _____

Problem Solving
Tessellations

Create a tessellation using the given figure.

1.

2.

Choose the letter for the best answer.

3. Which figure will NOT make a tessellation?

A

C

B

D

4. Which nonregular polygon can always be used to tile a floor?

   F pentagon

   G triangle

   H octagon

   J hexagon

5. For a combination of regular polygons to tessellate, the angles that meet at each vertex must add to what?

   A 90°

   B 180°

   C 360°

   D 720°

Holt Mathematics

Name _____ Date _____ Class _____

Problem Solving
Perimeter and Area of Rectangles and Parallelograms

Use the following for Exercises 1–2. A quilt for a twin bed is 68 in. by 90 in.

1. What is the area of the backing applied to the quilt?

2. A ruffle is sewn to the edge of the quilt. How many feet of ruffle are needed to go all the way around the edge of the quilt?

_____

Use the following for Exercises 3–4. Jaime is building a rectangular dog run that is 12 ft by 8 ft.

3. If the run is cemented, how many square feet will be covered by cement?

4. How much fencing will be required to enclose the dog run?

_____

Use the following for Exercises 5–6. Jackie is painting the walls in a room. Two walls are 12 ft by 8 ft, and two walls are 10 ft by 8 ft. Choose the letter for the best answer.

5. What is the area of the walls to be painted?

   A 352 ft^2 **C** 704 ft^2

   B 176 ft^2 **D** 400 ft^2

6. If a can of paint covers 300 square feet, how many cans of paint should Jackie buy?

   F 1   **H** 3

   G 2   **J** 4

Use the following for Exercises 7–8. One kind of pool cover is a tarp that stretches over the area of the pool and is tied down on the edge of the pool. The cover extends 6 inches beyond the edge of the pool. Choose the letter for the best answer.

7. A rectangular pool is 20 ft by 10 ft. What is the area of the tarp that will cover the pool?

   A 200 ft^2 **C** 60 ft^2

   B 231 ft^2 **D** 215.25 ft^2

8. If the tarp costs $2.50 per square foot, how much will the tarp cost?

   F $500.00   **H** $150.00

   G $538.13   **J** $577.50

Holt Mathematics

Name _____ Date _____ Class _____

Write the correct answer.

1. Find the area of the material required to cover the kite pictured below.

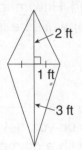

2. Find the area of the material required to cover the kite pictured below.

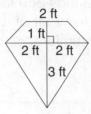

3. Find the approximate area of the state of Nevada.

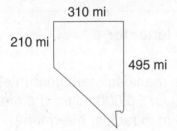

4. Find the area of the hexagonal gazebo floor.

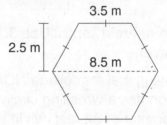

Choose the letter for the best answer.

5. Find the amount of flooring needed to cover the stage pictured below.

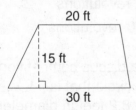

A 4500 ft²

B 750 ft²

C 525 ft²

D 375 ft²

6. Find the combined area of the congruent triangular gables.

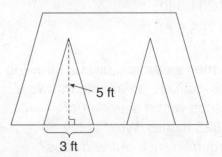

F 7.5 ft²

G 15 ft²

J 60 ft²

H 30 ft²

Holt Mathematics

Name _____ Date _____ Class _____

LESSON 8-3 **Problem Solving**
Circles

Round to the nearest tenth. Use 3.14 for π. Write the correct answer.

1. The world's tallest Ferris wheel is in Osaka, Japan, and stands 369 feet tall. Its wheel has a diameter of 328 feet. Find the circumference of the Ferris wheel.

   _____

2. A dog is on a 15-foot chain that is anchored to the ground. How much area can the dog cover while he is on the chain?

   _____

3. A small pizza has a diameter of 10 inches, and a medium has a diameter of 12 inches. How much more pizza do you get with the medium pizza?

   _____

4. How much more crust do you get with a medium pizza with a diameter of 12 inches than a small pizza with a 10 inch diameter?

   _____

Round to the nearest tenth. Use 3.14 for π. Choose the letter for the best answer.

5. The wrestling mat for college NCAA competition has a wrestling circle with a diameter of 32 feet, while a high school mat has a diameter of 28 feet. How much more area is there in a college wrestling mat than a high school mat?

   A 12.6 ft^2
   B 188.4 ft^2
   C 234.8 ft^2
   D 753.6 ft^2

6. Many tire manufacturers guarantee their tires for 50,000 miles. If a tire has a 16-inch radius, how many revolutions of the tire are guaranteed? There are 63,360 inches in a mile. Round to the nearest revolution.

   F 630.6 revolutions
   G 3125 revolutions
   H 31,528,662 revolutions
   J 500,000,000 revolutions

7. In men's Olympic discus throwing competition, an athlete throws a discus with a diameter of 8.625 inches. What is the circumference of the discus?

   A 13.5 in.
   B 27.1 in.
   C 58.4 in.
   D 233.6 in.

8. An athlete in a discus competition throws from a circle that is approximately 8.2 feet in diameter. What is the area of the discus throwing circle?

   F 52.8 ft^2
   G 25.7 ft^2
   H 12.9 ft^2
   J 211.1 ft^2

Copyright © by Holt, Rinehart and Winston.
All rights reserved.

PS58

Holt Mathematics

LESSON 8-4 Problem Solving
Drawing Three-Dimensional Figures

Write the correct answer.

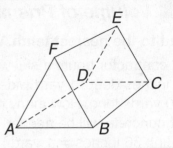

1. Mitch used a triangular prism in his science experiment. Name the vertices, edges, and faces of the triangular prism shown at the right.

_____

_____

_____

2. Amber used cubes to make the model shown below of a sculpture she wants to make. Draw the front, top, and side views of the model.

Choose the letter of the best answer.

3. Which is the front view of the figure shown at the left? ____

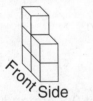

          A   **B**   **C**   **D**

4. Which is **not** an edge of the figure shown at the right?

  F $\overline{AB}$ **H** $\overline{EF}$

  G $\overline{EH}$ **J** $\overline{BD}$

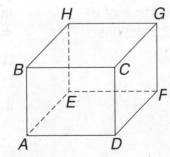

Holt Mathematics

Name _____ Date _____ Class _____

Problem Solving
Volume of Prisms and Cylinders

Round to the nearest tenth. Write the correct answer.

1. A contractor pours a sidewalk that is
   4 inches deep, 1 yard wide, and
   20 yards long. How many cubic yards
   of concrete will be needed?
   (Hint: 36 inches = 1 yard.)

   _____

2. A refrigerator has inside
   measurements of 50 cm by 118 cm
   by 44 cm. What is the capacity of the
   refrigerator?

   _____

**A rectangular box is 2 inches high, 3.5 inches wide and
4 inches long. A cylindrical box is 3.5 inches high and has
a diameter of 3.2 inches. Use 3.14 for π. Round to the nearest
tenth.**

3. Which box has a larger volume?

   _____

4. How much bigger is the larger box?

   _____

Use 3.14 for π. Choose the letter for the best answer.

5. A child's wading pool has a diameter
   of 5 feet and a height of 1 foot. How
   much water would it take to fill the
   pool? Round to the nearest gallon.
   (Hint: 1 cubic foot of water is
   approximately 7.5 gallons.)
   A 79 gallons
   B 589 gallons
   C 59 gallons
   D 147 gallons

6. How many cubic feet of air are in a
   room that is 15 feet long, 10 feet
   wide and 8 feet high?
   F 33 ft^3
   G 1200 ft^3
   H 1500 ft^3
   J 3768 ft^3

7. How many gallons of water will the
   water trough hold? Round to the
   nearest gallon. (Hint: 1 cubic foot of
   water is approximately 7.5 gallons.)

   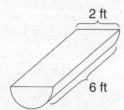
   2 ft
   6 ft

   A 19 gallons   **C** 141 gallons
   B 71 gallons   **D** 565 gallons

8. A can has diameter of 9.8 cm and is
   13.2 cm tall. What is the capacity of
   the can? Round to the nearest tenth.
   F 203.1 cm^3
   G 995.2 cm^3
   H 3980.7 cm^3
   J 959.2 cm^3

Holt Mathematics

LESSON 8-6 — Problem Solving
Volume of Pyramids and Cones

Round to the nearest tenth. Use 3.14 for π. Write the correct answer.

1. The Feathered Serpent Pyramid in Teotihuacan, Mexico is the third largest in the city. Its base is a square that measures 65 m on each side. The pyramid is 19.4 m high. What is the volume of the Feathered Serpent Pyramid?

2. The Sun Pyramid in Teotihuacan, Mexico, is larger than the Feathered Serpent Pyramid. The sides of the square base and the height are each about 3.3 times larger than the Feathered Serpent Pyramid. How many times larger is the volume of the Sun Pyramid than the Feathered Serpent Pyramid?

3. An oil funnel is in the shape of a cone. It has a diameter of 4 inches and a height of 6 inches. If the end of the funnel is plugged, how much oil can the funnel hold before it overflows?

4. One quart of oil has a volume of approximately 57.6 in^3. Does the oil funnel in exercise 3 hold more or less than 1 quart of oil?

Round to the nearest tenth. Use 3.14 for π. Choose the letter for the best answer.

5. An ice cream cone has a diameter of 4.2 cm and a height of 11.5 cm. What is the volume of the cone?

   A 18.7 cm^3

   B 25.3 cm^3

   C 53.1 cm^3

   D 212.3 cm^3

6. When decorating a cake, the frosting is put into a cone shaped bag and then squeezed out a hole at the tip of the cone. How much frosting is in a bag that has a radius of 1.5 inches and a height of 8.5 inches?

   F 5.0 in^3 H 15.2 in^3

   G 13.3 in^3 J 20.0 in^3

7. What is the volume of the hourglass at the right?

   A 13.1 in^3

   B 26.2 in^3

   C 52.3 in^3

   D 102.8 in^3

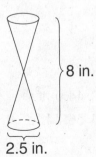

8 in.

2.5 in.

Holt Mathematics

Name _____ Date _____ Class _____

LESSON **Problem Solving**
8-7 *Surface Area of Prisms and Cylinders*

An important factor in designing packaging for a product is the
amount of material required to make the package. Consider the
three figures described in the table below. Use 3.14 for π.
Round to the nearest tenth. Write the correct answer.

1. Find the surface area of each
   package given in the table.

2. Which package has the lowest
   materials cost? Assume all of the
   packages are made from the same
   material.

_____

Package	Dimensions	Volume	Surface Area
Prism	Base: 2" × 16" Height = 2"	64 in³	
Prism	Base: 4" × 4" Height = 4"	64 in³	
Cylinder	Radius = 2" Height = 5.1"	64.06 in³	

Use 3.14 for π. Round to the nearest hundredth.

3. How much cardboard material is
   required to make a cylindrical
   oatmeal container that has a
   diameter of 12.5 cm and a height of
   24 cm, assuming there is no
   overlap? The container will have a
   plastic lid.

_____

4. What is the surface area of
   a rectangular prism that is
   5 feet by 6 feet by 10 feet?

_____

**Use 3.14 for π. Round to the nearest tenth. Choose the letter for
the best answer.**

5. How much metal is required to make
   the trough pictured below?

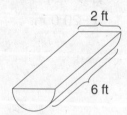

   2 ft

   6 ft

   A 22.0 ft²   C 44.0 ft²
   B 34.0 ft²   D 56.7 ft²

6. A can of vegetables has a diameter
   of 9.8 cm and is 13.2 cm tall. How
   much paper is required to make the
   label, assuming there is no overlap?
   Round to the nearest tenth.
   F 203.1 cm²
   G 406.2 cm²
   H 557.0 cm²
   J 812.4 cm²

Holt Mathematics

Name _____ Date _____ Class _____

Problem Solving

Surface Area of Pyramids and Cones

Round to the nearest tenth. Use 3.14 for π. Write the correct answer.

1. The Feathered Serpent Pyramid in Teotihuacan, Mexico, is the third largest in the city. Its base is a square that measures 65 m on each side. The pyramid is 19.4 m high and has a slant height of 37.8 m. The lateral faces of the pyramid are decorated with paintings. What is the surface area of the painted faces?

2. The Sun Pyramid in Teotihuacan, Mexico, is larger than the Feathered Serpent Pyramid. The sides of the square base and the slant height are each about 3.3 times larger than the Feathered Serpent Pyramid. How many times larger is the surface area of the lateral faces of the Sun Pyramid than the Feathered Serpent Pyramid?

3. An oil funnel is in the shape of a cone. It has a diameter of 4 inches and a slant height of 6 inches. How much material does it take to make a funnel with these dimensions?

4. If the diameter of the funnel in Exercise 6 is doubled, by how much does it increase the surface area of the funnel?

Round to the nearest tenth. Use 3.14 for π. Choose the letter for the best answer.

5. An ice cream cone has a diameter of 4.2 cm and a slant height of 11.5 cm. What is the surface area of the ice cream cone?

    A 4.7 cm^2 **C** 75.83 cm^2

    B 19.9 cm^2 **D** 159.2 cm^2

6. A marker has a conical tip. The diameter of the tip is 1 cm and the slant height is 0.7 cm. What is the area of the writing surface of the marker tip?

    F 1.1 cm^2 **H** 2.2 cm^2

    G 1.9 cm^2 **J** 5.3 cm^2

7. A skylight is shaped like a square pyramid. Each panel has a 4 m base. The slant height is 2 m, and the base is open. The installation cost is $5.25 per square meter. What is the cost to install 4 skylights?

    A $64   **C** $218

    B $159   **D** $336

8. A paper drinking cup shaped like a cone has a 10 cm slant height and an 8 cm diameter. What is the surface area of the cone?

    F 88.9 cm^2 **H** 251.2 cm^2

    G 125.6 cm^2 **J** 301.2 cm^2

Holt Mathematics

Name _____ Date _____ Class _____

Problem Solving
Spheres

Early golf balls were smooth spheres. Later it was discovered that golf balls flew better when they were dimpled. On January 1, 1932, the United States Golf Association set standards for the weight and size of a golf ball. The minimum diameter of a regulation golf ball is 1.680 inches. Use 3.14 for π. Round to the nearest hundredth.

1. Find the volume of a smooth golf ball with the minimum diameter allowed by the United States Golf Association.

   _____

2. Find the surface area of a smooth golf ball with the minimum diameter allowed by the United States Golf Association.

   _____

3. Would the dimples on a golf ball increase or decrease the volume of the ball?

   _____

4. Would the dimples on a golf ball increase or decrease the surface area of the ball?

   _____

Use 3.14 for π. Use the following information for Exercises 5–6. A track and field expert recommends changes to the size of a shot put. One recommendation is that a shot put should have a diameter between 90 and 110 mm. Choose the letter for the best answer.

5. Find the surface area of a shot put with a diameter of 90 mm.

   A 25,434 mm^2

   B 101,736 mm^2

   C 381,520 mm^2

   D 3,052,080 mm^2

6. Find the surface area of a shot put with diameter 110 mm.

   F 9,499 mm^2

   G 22,834 mm^2

   H 37,994 mm^2

   J 151,976 mm^2

7. Find the volume of the earth if the average diameter of the earth is 7926 miles.

   A 2.0×10^8 mi^3

   B 2.6×10^{11} mi^3

   C 7.9×10^8 mi^3

   D 2.1×10^{12} mi^3

8. An ice cream cone has a diameter of 4.2 cm and a height of 11.5 cm. One spherical scoop of ice cream is put on the cone that has a diameter of 5.6 cm. If the ice cream were to melt in the cone, how much of it would overflow the cone? Round to the nearest tenth.

   F 0 cm^3 H 38.8 cm^3

   G 12.3 cm^3 J 54.3 cm^3

Holt Mathematics

Name _____ Date _____ Class _____

Problem Solving
8-10 *Scaling Three-Dimensional Figures*

Round to the nearest hundredth. Write the correct answer.

1. The smallest regulation golf ball has a volume of 2.48 cubic inches. If the diameter of the ball were increased by 10%, or a factor of 1.1, what will the volume of the golf ball be?

2. The smallest regulation golf ball has a surface area of 8.86 square inches. If the diameter of the ball were increased by 10%, what will the surface area of the golf ball be?

3. The Feathered Serpent Pyramid in Teotihuacan, Mexico, is the third largest in the city. The dimensions of the Sun Pyramid in Teotihuacan, Mexico, are about 3.3 times larger than the Feathered Serpent Pyramid. How many times larger is the volume of the Sun Pyramid than the Feathered Serpent Pyramid?

4. The faces of the Feathered Serpent Pyramid and the Sun Pyramid described in Exercise 3 have ancient paintings on them. How many times larger is the surface area of the faces of the Sun Pyramid than the faces of the Feathered Serpent Pyramid?

Choose the letter for the best answer.

5. John is designing a shipping container that boxes will be packed into. The container he designed will hold 24 boxes. If he doubles the sides of his container, how many times more boxes will the shipping container hold?

  A 2   **C** 8
  B 4   **D** 192

6. If John doubles the sides of his container from exercise 5, how many times more material will be required to make the container?

  F 2   **H** 8
  G 4   **J** 192

7. A child's sandbox is shaped like a rectangular prism and holds 2 cubic feet of sand. The dimensions of the next size sandbox are double the smaller sandbox. How much sand will the larger sandbox hold?

  A 4 ft^3 **C** 16 ft^3
  B 8 ft^3 **D** 32 ft^3

8. Maria used two boxes of sugar cubes to create a solid building for a class project. She decides that the building is too small and she will rebuild it 3 times larger. How many more boxes of sugar cubes will she need?

  F 4   **H** 27
  G 25   **J** 52

Copyright © by Holt, Rinehart and Winston.
All rights reserved.

PS65

Holt Mathematics

Problem Solving

Samples and Surveys

Identify the sampling method used.

1. Every twentieth student on a list is chosen to participate in a poll.

2. Seat numbers are drawn from a hat to identify passengers on an airplane that will be surveyed.

Give a reason why the sample could be biased.

3. A company wants to find out how its customers rate their products. They ask people who visit the company's Web Site to rate their products.

4. A teacher polls all of the students who are in detention on Friday about their opinions on the amount of homework students should have each night.

A car dealership wants to know how people who have visited the dealership feel about the dealership and the sales people. They survey every 5th person who buys a car. Choose the letter for the best answer.

5. Identify the population.
   A People who visit the dealership
   B People who buy a car from the dealership
   C People in the local area
   D The salesmen at the dealership

6. Identify the sample.
   F Every person who visits the dealership
   G People who buy a car
   H Every 5th buyer
   J People in the local area

7. Identify the possible bias.
   A Not all people will visit the dealership.
   B Did not survey everyone who buys a car.
   C Not including those who visited but did not buy.
   D There is no bias.

8. Identify the sampling method used.
   F Random
   G Systematic
   H Stratified
   J None of these

Holt Mathematics

LESSON 9-2

Problem Solving

Organizing Data

A consumer survey gathered the following data about what teens do while on online.

1. Make a stem-and-leaf plot of the data.

Teens' Activities Online	
Activity	Percent
E-mail	95
Use search engines	86
Instant Messaging	82
Visit music sites	73
Enter contests	73

The stem-and-leaf plot that shows the total number of medals won by different countries in the 2000 Summer Olympics. Choose the letter for the best answer.

2. List all the data values in the stem-and-leaf plot.

   A 2, 4, 5, 6, 7, 8, 9

   B 23, 25, 26, 28, 28, 29, 34, 38, 40, 57, 58, 59, 60, 70, 88, 97

   C 23, 25, 26, 28, 29, 34, 38, 57, 58, 59, 88, 97

   D 23, 25, 26, 28, 28, 29, 34, 38, 57, 58, 59, 88, 97

2000 Olympic Medals

```
2 | 3 5 6 8 8 9
3 | 4 8
4 |
5 | 7 8 9
6 |
7 |
8 | 8
9 | 7
```

3. What is the least number of medals won by a country represented in the stem-and-leaf plot?

   F 3
   G 4
   H 23
   J 97

4. What is the greatest number of medals won by a country represented in the stem-and-leaf plot?

   A 9
   B 70
   C 79
   D 97

Holt Mathematics

LESSON 9-3

Problem Solving
Measures of Central Tendency

Use the data to find each answer.

1. Find the average number of passengers in the world's five busiest airports.

   _____

2. Find the median number of passengers in the world's five busiest airports.

   _____

World's Busiest Airports

Airport	Total Passengers (in millions)
Atlanta, Hartsfield	80.2
Chicago, O'Hare	72.1
Los Angeles	68.5
London, Heathrow	64.6
Dallas/Ft. Worth	60.7

3. Find the mode of the airport data.

   _____

4. Find the range of the airport data.

Choose the letter for the best answer.

5. What was the mean production of motor vehicles in 1998?

   A 8,651,500 vehicles

   B 10,249,250 vehicles

   C 11,264,250 vehicles

   D 12,000,000 vehicles

World Motor Vehicle Production (in thousands) 1998–1999

Country	1998	1999
United States	12,047	13,063
Canada	2,568	3,026
Europe	16,332	16,546
Japan	10,050	9904

6. What was the range of production in 1999?

   F 9,800,000 vehicles

   G 11,480,000 vehicles

   H 12,520,000 vehicles

   J 13,520,000 vehicles

7. What was the median number of vehicles produced in 1999?

   A 3,026,000 vehicles

   B 3,069,000 vehicles

   C 11,483,500 vehicles

   D 13,063,000 vehicles

8. Which value is largest?

   F Mean of 1998 data

   G Mean of 1999 data

   H Median of 1998 data

   J Median of 1999 data

Holt Mathematics

Name _____ Date _____ Class _____

Write the correct answer.

1. Find the median of the data.

_____

2. Find the first and third quartiles of the data.

_____

3. Make a box-and-whisker plot of the data.

Super Bowl Point Differences

Year	Point Difference
2001	27
2000	7
1999	15
1998	7
1997	14
1996	10
1995	23
1994	17
1993	35
1992	13

The box-and-whisker plots compare the highest recorded Fahrenheit temperatures on the seven continents with the lowest recorded temperatures. Choose the letter for the best answer.

4. Which statement is true?

   A The median of the high temperatures is less than the median of the low temperatures.

   B The range of low temperatures is greater than the range of high temperatures.

   C The range of the middle half of the data is greater for the high temperatures.

   D The median of the high temperatures is 49°F.

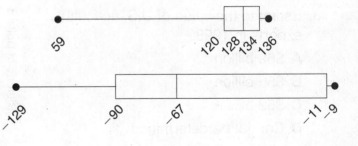

5. What is the median of the high temperatures?

   F 128°F   H −67°F

   G 120°F   J −90°F

6. What is the range of the low temperatures?

   A 77°F   C 120°F

   B 79°F   D 129°F

Holt Mathematics

Name _____ Date _____ Class _____

Problem Solving
Displaying Data

Make the indicated graph.

1. Make a double-bar graph of the homework data.

Hours of Daily Homework	1	2	3	4	5
Boys	12	5	2	1	0
Girls	4	6	5	3	2

2. The annual hourly delay per driver in the 20 U.S. cities with the most traffic are as follows: 56, 42, 53, 46, 34, 37, 42, 34, 53, 21, 45, 50, 34, 42, 41, 38, 42, 34, 38, 31. Make a histogram with intervals of 5 hours.

For 3–5, refer to the double-line graph.
Circle the letter of the correct answer.

3. Estimate the value of U.S. agricultural exports in 1998.

   A $62 billion

   B $59 billion

   C $52 billion

   D Cannot be determined

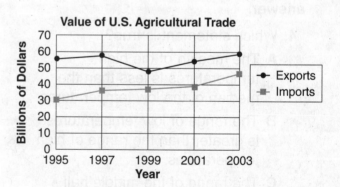

4. Estimate the value of U.S. agricultural imports in 2000.

   F $39 billion

   G $31 billion

   H $29 billion

   J $21 billion

5. Estimate the difference between agricultural exports and imports in 1995.

   A $16 billion

   B $21 billion

   C $26 billion

   D Cannot be determined

Holt Mathematics

Problem Solving

LESSON 9-6

Misleading Graphs and Statistics

Explain why each statistics is misleading.

1. A poll taken at a college says that 38% of students like pizza the best, 32% like hamburgers the best, and 30% like spaghetti the best. They conclude that most of the students at the college like pizza the best.

_____

_____

2. The National Safety Council of Ireland found that young men were responsible in 57% of automobile accidents they were involved in. The NSC Web site made this claim: "Young men are responsible for over half of all road accidents."

_____

_____

3. Explain why the Centers for Disease Control (CDC) has been highly criticized for the graph below.

_____

_____

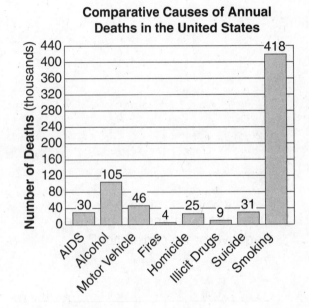

Comparative Causes of Annual Deaths in the United States

Choose the letter for the best answer.

4. Which statement is a misleading statistic for the data in the table?

Student	Test Grade
A	85%
B	92%
C	88%
D	10%
E	80%

A The median score was 85%.

B Most students scored an 80% or above.

C The average test score was 71%.

D The range of test scores was 82.

5. A sno-cone store claims, "Our sales have tripled!" Sno-cone sales from March to May were 50 and sales from June to August were 150. Why is this misleading?

A Sample size is too small.

B During the summer, sales should be higher.

C Should use the median not mean.

D The statement isn't misleading.

Holt Mathematics

LESSON 9-7 Problem Solving
Scatter Plots

Use the data given at the right.

1. Make a scatter plot of the data.

Percent of Americans Who Have Completed High School

Year	Percent
1910	13.5
1920	16.4
1930	19.1
1940	24.5
1950	34.3
1960	41.1
1970	55.2
1980	68.6
1990	77.6
1999	83.4

2. Does the data show a positive, negative or no correlation?

3. Use the scatter plot to predict the percent of Americans who will complete high school in 2010.

_____

Choose the letter for the best answer.

4. Which data sets have a positive correlation?

   A The length of the lines at amusement park rides and the number of rides you can ride in a day

   B The temperature on a summer day and the number of visitors at a swimming pool

   C The square miles of a state and the population of the state in the 2000 census

   D The length of time spent studying and doing homework and the length of time spent doing other activities

5. Which data sets have a negative correlation?

   F The number of visitors at an amusement park and the length of the lines for the rides

   G The amount of speed over the speed limit when you get a speeding ticket and the amount of the fine for speeding

   H The temperature and the number of people wearing coats

   J The distance you live from school and the amount of time it takes to get to school

Holt Mathematics

Problem Solving

LESSON 9-8

Choosing the Best Representation of Data

Write what kind of graph would be best to display the described data.

1. Numbers of times that members of track team ran a mile in the following intervals: 4 min 31 s to 4 min 40 s, 4 min 41 s to 4 min 50 s, 4 min 51 s to 5 min, 5 min 1 s to 5 min 10 s

_____

2. Distribution and range of students' scores on a history exam

_____

3. Relationship between the amounts of time a student spent on her math homework and the numbers of homework problems she solved

_____

4. Total numbers of victories of eight teams in an intramural volleyball league

_____

5. Part of calories in a meal that come from protein

_____

6. Numbers of books that a student reads each month over a year

_____

Choose the letter for the best answer.

7. A bar graph is a good way to display
   - **A** data that changes over time.
   - **B** parts of a whole.
   - **C** distribution of data.
   - **D** comparison of different groups of data.

8. A circle graph is a good way to display
   - **F** range and distribution of data.
   - **G** distribution of data.
   - **H** parts of a whole.
   - **J** changes in data over time.

9. A scatter plot is a good way to display
   - **A** comparison of different groups of data.
   - **B** distribution and range of data.
   - **C** the relationship between two sets of data.
   - **D** parts of a whole.

10. A box-and-whisker plot is a good way to display
   - **F** range and distribution of data.
   - **G** the relationship between two sets of data.
   - **H** data that changes over time.
   - **J** parts of a whole.

Holt Mathematics

LESSON	**Problem Solving**
10-1	*Probability*

Write the correct answer.

1. To get people to buy more of their product, a company advertises that in selected boxes of their popsicles is a super hero trading card. There is a $\frac{1}{4}$ chance of getting a trading card in a box. What is the probability that there will not be a trading card in the box of popsicles that you buy?

2. The probability of winning a lucky wheel television game show in which 6 preselected numbers are spun on a wheel numbered 1−49 is $\frac{1}{13,983,816}$ or 0.000007151%. What is the probability that you will not win the game show?

Based on world statistics, the probability of identical twins is 0.004, while the probability of fraternal twins is 0.023.

3. What is the probability that a person chosen at random from the world will be a twin?

4. What is the probability that a person chosen at random from the world will not be a twin?

Use the table below that shows the probability of multiple births by country. Choose the letter for the best answer.

5. In which country is it most likely to have multiple births?

   A Japan   C Sweden

   B United States   D Switzerland

6. In which country is it least likely to have multiple births?

   F Japan   H Sweden

   G United States   J Switzerland

7. In which two countries are multiple births equally likely?

   A United Kingdom, Canada

   B Canada, Switzerland

   C Sweden, United Kingdom

   D Japan, United States

Probability of Multiple Births

Country	Probability
Canada	0.012
Japan	0.008
United Kingdom	0.014
United States	0.029
Sweden	0.014
Switzerland	0.013

Holt Mathematics

Name _____ Date _____ Class _____

Use the table below. Round to the nearest percent. Write the correct answer.

Average Number of Days of Sunshine Per Year for Selected Cities

City	Number of Days
Buffalo, NY	175
Fort Wayne, IN	215
Miami, FL	256
Raleigh, NC	212
Richmond, VA	230

1. Estimate the probability of sunshine in Buffalo, NY.

   _____

2. Estimate the probability of sunshine in Fort Wayne, IN.

   _____

3. Estimate the probability of sunshine in Miami, FL.

   _____

4. Estimate the probability that it will not be sunny in Raleigh, NC.

   _____

5. Estimate the probability that it will not be sunny in Miami, FL.

   _____

6. Estimate the probability of sunshine in Richmond, VA.

   _____

Use the table below that shows the number of deaths and injuries caused by lightning strikes. Choose the letter for the best answer.

States with Most Lightning Deaths

State	Average deaths per year	Average injuries per year	Population
Florida	9.6	32.7	15,982,378
North Carolina	4.6	12.9	8,049,313
Texas	4.6	9.3	20,851,820
New York	3.6	12.5	18,976,457
Tennessee	3.4	9.7	5,689,283

7. Estimate the probability of being injured by a lightning strike in New York.

   A 0.0000007%   C 0.00007%

   B 0.0000002%   D 0.000002%

8. Estimate the probability of being killed by lightning in North Carolina.

   F 0.0000006%   H 0.00002%

   G 0.00006%   J 0.000002%

9. Estimate the probability of being struck by lightning in Florida.

   A 0.00006%

   B 0.00026%

   C 0.0000026%

   D 0.0006%

10. In which two states do you have the highest probability of being struck by lightning?

   F Florida, North Carolina

   G Florida, Tennessee

   H Texas, New York

   J North Carolina, Tennessee

Holt Mathematics

LESSON 10-3 Problem Solving
Use a Simulation

Use the table of random numbers below. Use at least 10 trials to simulate each situation. Write the correct answer.

1. Of people 18–24 years of age, 49% do volunteer work. If 10 people ages 18–24 were chosen at random, estimate the probability that at least 4 of them do volunteer work.

87244	11632	85815	61766
19579	28186	18533	24633
74581	65633	54238	32848
87549	85976	13355	46498
53736	21616	86318	77291
24794	31119	48193	44869
86585	27919	65264	93557
94425	13325	16635	25840
18394	73266	67899	38783
94228	23426	76679	41256

_____

2. In the 2000 Presidential election, 56% of the population of North Carolina voted for George W. Bush. If 10 people were chosen at random from North Carolina, estimate the probability that at least 8 of them voted for Bush.

_____

3. Forty percent of households with televisions watched the 2001 Super Bowl game. If 10 households with televisions are chosen at random, estimate the probability that at least 3 watched the 2001 Super Bowl.

Use the table above and at least 10 trials to simulate each situation. Choose the letter for the best estimate.

4. As of August 2000, 42% of U.S. households had Internet access. If 10 households are chosen at random, estimate the probability that at least 5 of them will have Internet access.

A 0%   C 60%

B 30%   D 90%

5. On average, there is rain 20% of the days in April in Orlando, FL. Estimate the probability that it will rain at least once during your 7-day vacation in Orlando in April.

F 20%   H 70%

G 50%   J 40%

6. Kareem Abdul-Jabaar is the NBA lifetime leader in field goals. During his career, he made 56% of the field goals he attempted. In a given game, estimate the probability that he would make at least 6 out of 10 field goals.

A 40%   C 80%

B 60%   D 100%

7. At the University of Virginia 39% of the applicants are accepted. If 10 applicants to the University of Virginia are chosen at random, estimate the probability that at least 4 of them are accepted to the University of Virginia.

F 10%   H 80%

G 40%   J 70%

Holt Mathematics

Name _____ Date _____ Class _____

Problem Solving
Theoretical Probability

A company that sells frozen pizzas is running a promotional special. Out of the next 100,000 boxes of pizza produced, randomly chosen boxes will be prize winners. There will be one grand prize winner who will receive $100,000. Five hundred first prize winners will get $1000, and 3,000 second prize winners will get a free pizza. Write the correct answer in fraction and percent form.

1. What is the probability that the box of pizza you just bought will be a grand prize winner?

_____

2. What is the probability that the box of pizza you just bought will be a first prize winner?

_____

3. What is the probability that the box of pizza you just bought will be a second prize winner?

_____

4. What is the probability that you will win anything with the box of pizza you just bought?

_____

Researchers at the National Institutes of Health are recommending that instead of screening all people for certain diseases, they can use a Punnett square to identify the people who are most likely to have the disease. By only screening these people, the cost of screening will be less. Fill in the Punnett square below and use them to choose the letter for the best answer.

5. What is the probability of DD?

    A 0%   **C** 50%

    B 25%   **D** 75%

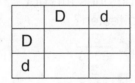

	D	d
D		
d		

6. What is the probability of Dd?

    F 25%   **H** 75%

    G 50%   **J** 100%

8. DD or Dd indicates that the patient will have the disease. What is the probability that the patient will have the disease?

7. What is the probability of dd?

    A 0%   **C** 50%

    B 25%   **D** 75%

    F 25%   **H** 75%

    G 50%   **J** 100%

Holt Mathematics

Name _____ Date _____ Class _____

Problem Solving
Independent and Dependent Events

Are the events independent or dependent? Write the correct answer.

1. Selecting a piece of fruit, then choosing a drink.

2. Buying a CD, then going to another store to buy a video tape if you have enough money left.

_____

_____

Dr. Fred Hoppe of McMaster University claims that the probability of winning a pick 6 number game where six numbers are drawn from the set 1 through 49 is about the same as getting 24 heads in a row when you flip a fair coin.

3. Find the probability of winning the pick 6 game and the probability of getting 24 heads in a row when you flip a fair coin.

4. Which is more likely: to win a pick 6 game or to get 24 heads in a row when you flip a fair coin?

_____

_____

In a shipment of 20 computers, 3 are defective. Choose the letter for the best answer.

5. Three computers are randomly selected and tested. What is the probability that all three are defective if the first and second ones are not replaced after being tested?

   A $\frac{1}{760}$ C $\frac{27}{8000}$

   B $\frac{1}{1140}$ D $\frac{3}{5000}$

6. Three computers are randomly selected and tested. What is the probability that all three are defective if the first and second ones are replaced after being tested?

   F $\frac{1}{760}$ H $\frac{27}{8000}$

   G $\frac{1}{1140}$ J $\frac{3}{5000}$

7. Three computers are randomly selected and tested. What is the probability that none are defective if the first and second ones are not replaced after being tested?

   A $\frac{34}{57}$ C $\frac{4913}{6840}$

   B $\frac{4913}{8000}$ D $\frac{1}{2000}$

8. Three computers are randomly selected and tested. What is the probability that none are defective if the first and second ones are replaced after being tested?

   F $\frac{34}{57}$ H $\frac{4913}{6840}$

   G $\frac{4913}{8000}$ J $\frac{1}{2000}$

Holt Mathematics

Problem Solving
LESSON 10-6

Making Decisions and Predictions

Write the correct answer.

1. A quality control inspector at a light bulb factory finds 2 defective bulbs in a batch of 1000 light bulbs. If the plant manufactures 75,000 light bulbs in one day, predict how many will be defective.

2. A game consists of rolling two fair number cubes labeled 1–6. Add both numbers. Player A wins if the sum is greater than 10. Player B wins if the sum is 7. Is the game fair or not? Explain.

3. A spinner has 5 equal sections numbered 1–5. Predict how many times Kevin will spin an even number in 40 spins.

4. In her last six 100-meter runs, Lee had the following times in seconds: 12:04, 13:11, 12:25, 11:58, 12:37, and 13:20. Based on these results, what is the best prediction of the number of times Lee will run faster than 13 seconds in her next 30 runs?

Use the table below that shows the number of colors of the last 200 T-shirts sold at a T-shirt shop. The manager of the store wants to order 1800 new T-shirts. Choose the letter of the best answer

5. How many red T-shirts should the manager order?

   A 175   C 378

   B 315   D 630

6. How many blue T-shirts should the manager order?

   F 495   H 900

   G 665   J 990

7. How many more black T-shirts than white T-shirts should the manager order?

   A 855   C 315

   B 585   D 270

T-Shirts Sold

Color	Number
Red	35
Blue	55
Green	15
Black	65
White	30

Holt Mathematics

LESSON 10-7 Problem Solving

Odds

In the last 25 Summer Olympics since 1900, an American man has won the gold medal in the 400-meter dash 18 times. Write the correct answer.

1. Find the probability that an American man will win the gold medal in the 400-meter dash in the next Summer Olympics.

_____

2. Find the probability that an American man will not win the gold medal in the 400-meter dash in the next Summer Olympics.

_____

3. Find the odds that an American man will win the gold medal in the 400-meter dash in the next Summer Olympics.

_____

4. Find the odds that an American man will not win the gold medal in the 400-meter dash in the next Summer Olympics.

_____

Use the table below that shows the probability that a player will end up on a certain square after a single roll in a game of Monopoly.

5. What are the odds that you will end up in jail on your next roll in a game of Monopoly?

   A 39:1000

   B 39:961

   C 1000:961

   D 961:39

Probability of Ending Up on a Monopoly Square

Square	Probability	Rank
In Jail	$\frac{39}{1000}$	1
Illinois Ave.	$\frac{32}{1000}$	2
Go	$\frac{31}{1000}$	3
Boardwalk	$\frac{26}{1000}$	18
Park Place	$\frac{22}{1000}$	33

6. What are the odds that you will end up on Boardwalk on your next roll in a game of Monopoly?

   A 13:500   C 13:487

   B 500:13   D 487:13

7. What are the odds that you will not end up on Boardwalk on your next roll in a game of Monopoly?

   F 487:500   H 13:487

   G 500:487   J 487:13

8. What are the odds that you will end up on Go on your next roll in a game of Monopoly?

   A 31:969   C 31:1000

   B 969:31   D 1000:31

9. What are the odds that you will not end up on Park Place on your next roll in a game of Monopoly?

   F 11:489   H 489:500

   G 489:11   J 500:489

Holt Mathematics

Name _____ Date _____ Class _____

Problem Solving
Counting Principles

Write the correct answer.

1. The 5-digit zip code system for United States mail was implemented in 1963. How many different possibilities of zip codes are there with a 5-digit zip code where each digit can be 0 through 9?

2. In 1983, the ZIP +4 zip code system was introduced so mail could be more easily sorted by the 5-digit zip code plus an additional 4 digits. How many different possibilities of zip codes are there with the ZIP +4 system?

3. In Canada, each postal code has 6 symbols. The first, third and fifth symbols are letters of the alphabet and the second, fourth and sixth symbols are digits from 0 through 9. How many possible postal codes are there in Canada?

4. In the United Kingdom the postal code has 6 symbols. The first, second, fifth and sixth are letters of the alphabet and the third and fourth are digits from 0 through 9. How many possible postal codes are there in the United Kingdom?

Choose the letter for the best answer.

5. In Sharon Springs, Kansas, all of the phone numbers begin 852–4. The only differences in the phone numbers are the last 3 digits. How many possible phone numbers can be assigned using this system?

   A 729   C 6561

   B 1000   D 10,000

6. Many large cities have run out of phone numbers and so a new area code must be introduced. How many different phone numbers are there in a single area code if the first digit can't be zero?

   F 90,000   H 9,000,000

   G 4,782,969   J 10,000,000

7. How many different phone numbers are possible using a 3-digit area code and a 7-digit phone number if the first digit of the area code and phone number cannot be zero?

   A 3,486,784,401   C 9,500,000,000

   B 8,100,000,000   D 10,000,000,000

8. A shipping service offers to send packages by ground delivery using 2 different companies, by next day air using 3 different companies, and by 2-day air using 3 different companies. How many different shipping options does the service offer?

   F 3   H 10

   G 8   J 18

Holt Mathematics

Name _____ Date _____ Class _____

Problem Solving
Permutations and Combinations

Write the correct answer.

1. In a day camp, 6 children are picked to be team captains from the group of children numbered 1 through 49. How many possibilities are there for who could be the 6 captains?

2. If you had to match 6 players in the correct order for most popular outfielder from a pool of professional players numbered 1 through 49, how many possibilities are there?

Volleyball tournaments often use pool play to determine which teams will play in the semi-final and championship games. The teams are divided into different pools, and each team must play every other team in the pool. The teams with the best record in pool play advance to the final games.

3. If 12 teams are divided into 2 pools, how many games will be played in each pool?

4. If 12 teams are divided into 3 pools, how many pool play games will be played in each pool?

A word jumble game gives you a certain number of letters that you must make into a word. Choose the letter for the best answer.

5. How many possibilities are there for a jumble with 4 letters?

   A 4   C 24
   B 12   D 30

6. How many possibilities are there for a jumble with 5 letters?

   F 24   H 120
   G 75   J 150

7. How many possibilities are there for a jumble with 6 letters?

   A 120
   B 500
   C 720
   D 1000

8. On the Internet, a site offers a program that will un-jumble letters and give you all of the possible words that can be made with those letters. However, the program will not allow you to enter more than 7 letters due to the amount of time it would take to analyze. How many more possibilities are there with 8 letters than with 7?

   F 5040   G 20,640
   H 35,280   J 40,320

Holt Mathematics

Name _____ Date _____ Class _____

Problem Solving
Simplifying Algebraic Expressions

Write the correct answer.

1. An item costs *x* dollars. The tax rate is 5% of the cost of the item, or 0.05*x*. Write and simplify an expression to find the total cost of the item with tax.

_____

2. A sweater costs *d* dollars at regular price. The sweater is reduced by 20%, or 0.2*d*. Write and simplify an expression to find the cost of the sweater before tax.

_____

3. Consecutive integers are integers that differ by one. You can represent consecutive integers as *x*, *x* + 1, *x* + 2 and so on. Write an equation and solve to find three consecutive integers whose sum is 33.

4. Consecutive even integers can be represented by *x*, *x* + 2, *x* + 4 and so on. Write an equation and solve to find three consecutive even integers whose sum is 54.

Choose the letter for the best answer.

5. In Super Bowl XXXV, the total number of points scored was 41. The winning team outscored the losing team by 27 points. What was the final score of the game?

A 33 to 8

B 34 to 7

C 22 to 2

D 18 to 6

6. A high school basketball court is 34 feet longer than it is wide. If the perimeter of the court is 268, what are the dimensions of the court?

F 234 ft by 34 ft

G 67 ft by 67 ft

H 70 ft by 36 ft

J 84 ft by 50 ft

7. Julia ordered 2 hamburgers and Steven ordered 3 hamburgers. If their total bill before tax was $7.50, how much did each hamburger cost?

A $1.50

B $1.25

C $1.15

D $1.02

8. On three tests, a student scored a total of 258 points. If the student improved his performance on each test by 5 points, what was the score on each test?

F 81, 86, 91

G 80, 85, 90

H 75, 80, 85

J 70, 75, 80

Holt Mathematics

Problem Solving

LESSON 11-2 *Solving Multi-Step Equations*

A taxi company charges $2.25 for the first mile and then $0.20
per mile for each mile after the first, or $F = \$2.25 + \$0.20(m - 1)$
where F is the fare and m is the number of miles.

1. If Juan's taxi fare was $6.05, how
many miles did he travel in the taxi?

2. If Juan's taxi fare was $7.65, how
many miles did he travel in the taxi?

_____

_____

A new car loses 20% of its original value when you buy it and
then 8% of its original value per year, or $D = 0.8V - 0.08Vy$
where D is the value after y years with an original value V.

3. If a vehicle that was valued at
$20,000 new is now worth $9,600,
how old is the car?

4. A 6-year old vehicle is worth
$12,000. What was the original value
of the car?

_____

_____

The equation used to estimate typing speed is $S = \frac{1}{5}(w - 10e)$,
where S is the accurate typing speed, w is the number of words
typed in 5 minutes and e is the number of errors. Choose the
letter of the best answer.

5. Jane can type 55 words per minute
(wpm). In 5 minutes, she types 285
words. How many errors would you
expect her to make?

   A 0   **C** 2

   B 1   **D** 5

6. If Alex types 300 words in 5 minutes
with 5 errors, what is his typing
speed?

   F 48 wpm   **H** 59 wpm

   G 50 wpm   **J** 60 wpm

7. Johanna receives a report that says
her typing speed is 65 words per
minute. She knows that she made
4 errors in the 5-minute test. How
many words did she type in 5
minutes?

   A 285   **C** 365

   B 329   **D** 1825

8. Cecil can type 35 words per minute.
In 5 minutes, she types 255 words.
How many errors would you expect
her to make?

   F 2   **H** 6

   G 4   **J** 8

Holt Mathematics

Problem Solving
LESSON 11-3

Solving Equations with Variables on Both Sides

**The chart below describes three long-distance calling plans.
Round to the nearest minute. Write the correct answer.**

1. For what number of minutes will plan A and plan B cost the same?

_____

2. For what number of minutes per month will plan B and plan C cost the same?

_____

Long-Distance Plans

Plan	Monthly Access Fee	Charge per minute
A	$3.95	$0.08
B	$8.95	$0.06
C	$0	$0.10

3. For what number of minutes will plan A and plan C cost the same?

_____

Choose the letter for the best answer.

4. Carpet Plus installs carpet for $100 plus $8 per square yard of carpet. Carpet World charges $75 for installation and $10 per square yard of carpet. Find the number of square yards of carpet for which the cost including carpet and installation is the same.

A 1.4 yd^2 **C** 12.5 yd^2

B 9.7 yd^2 **D** 87.5 yd^2

5. One shuttle service charges $10 for pickup and $0.10 per mile. The other shuttle service has no pickup fee but charges $0.35 per mile. Find the number of miles for which the cost of the shuttle services is the same.

F 2.5 miles

G 22 miles

H 40 miles

J 48 miles

6. Joshua can purchase tile at one store for $0.99 per tile, but he will have to rent a tile saw for $25. At another store he can buy tile for $1.50 per tile and borrow a tile saw for free. Find the number of tiles for which the cost is the same. Round to the nearest tile.

A 10 tiles   **C** 25 tiles

B 13 tiles   **D** 49 tiles

7. One plumber charges a fee of $75 per service call plus $15 per hour. Another plumber has no flat fee, but charges $25 per hour. Find the number of hours for which the cost of the two plumbers is the same.

F 2.1 hours   **H** 7.5 hours

G 7 hours   **J** 7.8 hours

Holt Mathematics

Problem Solving

LESSON 11-4 *Solving Inequalities by Multiplying or Dividing*

Write the correct answer

1. A bottle contains at least 4 times as much juice as a glass contains. The bottle contains 32 fluid ounces. Write an inequality that shows this relationship.

_____

2. Solve the inequality in Exercise 1. What is the greatest amount the glass could contain?

_____

3. In the triple jump, Katrina jumped less than one-third the distance that Paula jumped. Katrina jumped 5 ft 6 in. Write an inequality that shows this relationship.

_____

4. Solve the inequality in Exercise 3. How far could Paula could have jumped?

_____

Choose the letter for the best answer.

5. Melinda earned at least 3 times as much money this month as last month. She earned $567 this month. Which inequality shows this relationship?

   A $567 < x$ **C** $567 > 3x$

   B $567 < 3x$ **D** $567 \geq 3x$

6. The shallow end of a pool is less than one-quarter as deep as the deep end. The shallow end is 3 feet deep. Which inequality shows this relationship?

   F $4 > 3x$ **H** $\frac{x}{4} > 3$

   G $4x < 3$ **J** $\frac{x}{4} < 3$

7. Arthur worked in the garden more than half as long as his brother. Arthur worked 6 hours in the garden. How long did his brother work in the garden?

   A less than 3 hours

   B 3 hours

   C less than 12 hours

   D more than 12 hours

8. The distance from Bill's house to the library is no more than 5 times the distance from his house to the park. If Bill's house is 10 miles from the library, what is the greatest distance his house could be from the park?

   F 2 miles

   G more than 2 miles

   H 20 miles

   J less than 20 miles

Holt Mathematics

Problem Solving

11-5 Solving Two-Step Inequalities

A school club is selling printed T-shirts to raise $650 for a trip. The table shows the profit they will make on each shirt after they pay the cost of production.

1. Suppose the club already has $150, at least how many 50/50 shirts must they sell to make enough money for the trip?

Shirt	Profit
50/50	$5.50
100% cotton	$7.82

_____

2. Suppose the club already has $100, but it plans to spend $50 on advertising. At least how many 100% cotton shirts must they sell to make enough money for the trip?

3. Suppose the club sold thirty 50/50 shirts on the first day of sales. At least how many more 50/50 shirts must they sell to make enough money for the trip?

_____

For Exercises 4–5, use this equation to estimate typing speed, $S = \frac{w}{5} - 2e$, where S is the accurate typing speed, w is the number of words typed in 5 minutes, and e is the number of errors. Choose the letter for the best answer.

4. One of the qualifications for a job is a typing speed of at least 65 words per minute. If Jordan knows that she will be able to type 350 words in five minutes, what is the maximum number of errors she can make?

   A 0   **C** 3

   B 2   **D** 4

5. Tanner usually makes 3 errors every 5 minutes when he is typing. If his goal is an accurate typing speed of at least 55 words per minute, how many words does he have to be able to type in 5 minutes?

   F 61 words   **H** 305 words

   G 300 words   **J** 325 words

6. A taxi charges $2.05 per ride and $0.20 for each mile, which can be written as $F = \$2.05 + \$0.20m$. How many miles can you travel in the cab and have the fare be less than $10?

   A 15   **C** 39

   B 25   **D** 43

7. Celia's long distance company charges $5.95 per month plus $0.06 per minute. If Celia has budgeted $30 for long distance, what is the maximum number of minutes she can call long distance per month?

   F 375 minutes   **H** 405 minutes

   G 400 minutes   **J** 420 minutes

Holt Mathematics

LESSON 11-6 Problem Solving
Systems of Equations

After college, Julia is offered two different jobs. The table summarizes the pay offered with each job. Write the correct answer.

1. Write an equation that shows the pay y of Job A after x years.

   _____

Job	Yearly Salary	Yearly Increase
A	$20,000	$2500
B	$25,000	$2000

2. Write an equation that shows the pay y of Job B after x years.

   _____

3. Is (8, 35,000) a solution to the system of equations in Exercises 1 and 2?

4. Solve the system of equations in Exercises 1 and 2.

   _____

5. If Julia plans to stay at this job only a few years and pay is the only consideration, which job should she choose?

A travel agency is offering two Orlando trip plans that include hotel accommodations and pairs of tickets to theme parks. Use the table below. Choose the letter for the best answer.

6. Find an equation about trip A where x represents the hotel cost per night and y represents the cost per pair of theme park tickets.

   A $5x + 2y = 415$ **C** $8x + 6y = 415$
   B $2x + 3y = 415$ **D** $3x + 2y = 415$

Trip	Number of nights	Pairs of theme park tickets	Cost
A	3	2	$415
B	5	4	$725

7. Find an equation about trip B where x represents the hotel cost per night and y represents the cost per pair of theme park tickets.

   F $5x + 4y = 725$
   G $4x + 5y = 725$
   H $8x + 6y = 725$
   J $3x + 4y = 725$

8. Solve the system of equations to find the nightly hotel cost and the cost for each pair of theme park tickets.

   A ($50, $105)
   B ($125 $20)
   C ($105, $50)
   D ($115, $35)

Holt Mathematics

Name _____ Date _____ Class _____

Problem Solving
Graphing Linear Equations

Write the correct answer.

1. The distance in feet traveled by a falling object is found by the formula $d = 16t^2$ where d is the distance in feet and t is the time in seconds. Graph the equation. Is the equation linear?

_____

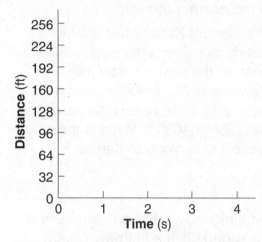

2. The formula that relates Celsius to Fahrenheit is $F = \frac{9}{5}C + 32$. Graph the equation. Is the equation linear?

_____

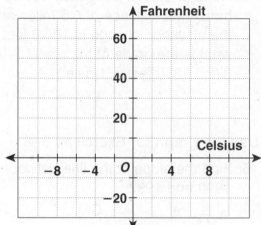

Wind chill is the temperature that the air feels like with the effect of the wind. The graph below shows the wind chill equation for a wind speed of 25 mph. For Exercises 3–6, refer to the graph.

3. If the temperature is 40° with a 25 mph wind, what is the wind chill?

   A 6°   **C** 29°
   B 20°   **D** 40°

4. If the temperature is 20° with a 25 mph wind, what is the wind chill?

   F 3°   **H** 13°
   G 10°   **J** 20°

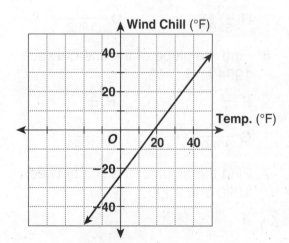

5. If the temperature is 0° with a 25 mph wind, what is the wind chill?

   A −30°   **C** −15°
   B −24°   **D** 0°

6. If the wind chill is 10° and there is a 25 mph wind, what is the actual temperature?

   F −11°   **H** 15°
   G 0°   **J** 25°

Holt Mathematics

Name _____ Date _____ Class _____

Problem Solving
Slope of a Line

Write the correct answer.

1. The state of Kansas has a fairly steady slope from the east to the west. At the eastern side, the elevation is 771 ft. At the western edge, 413 miles across the state, the elevation is 4039 ft. What is the approximate slope of Kansas?

2. The Feathered Serpent Pyramid in Teotihuacan, Mexico, has a square base. From the center of the base to the center of an edge of the pyramid is 32.5 m. The pyramid is 19.4 m high. What is the slope of each face of the pyramid?

3. On a highway, a 6% grade means a slope of 0.06. If a highway covers a horizontal distance of 0.5 miles and the elevation change is 184.8 feet, what is the grade of the road? (Hint: 5280 feet = 1 mile.)

4. The roof of a house rises vertically 3 feet for every 12 feet of horizontal distance. What is the slope, or pitch of the roof?

Use the graph for Exercises 5–8.

5. Find the slope of the line between 1990 and 1992.

   A $\frac{2}{11}$ **C** $\frac{11}{2}$

   B $\frac{35}{3982}$ **D** $\frac{11}{1992}$

6. Find the slope of the line between 1994 and 1996.

   F $\frac{7}{2}$ **H** $\frac{2}{7}$

   G $\frac{37}{3990}$ **J** $\frac{7}{1996}$

7. Find the slope of the line between 1998 and 2000.

   A 1

   B $\frac{1}{999}$

   C $\frac{1}{1000}$

   D 2

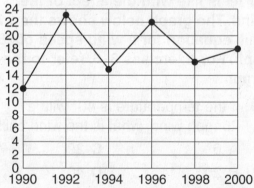

Number of Earthquakes Worldwide with a Magnitude of 7.0 or Greater

8. What does it mean when the slope is negative?

   F The number of earthquakes stayed the same.

   G The number of earthquakes increased.

   H The number of earthquakes decreased.

   J It means nothing.

Holt Mathematics

Name _____ Date _____ Class _____

Write the correct answer.

1. Jaime purchased a $20 bus pass. Each time she rides the bus, $1.25 is deducted from the pass. The linear equation $y = -1.25x + 20$ represents the amount of money on the bus pass after x rides. Identify the slope and the x- and y-intercepts. Graph the equation at the right.

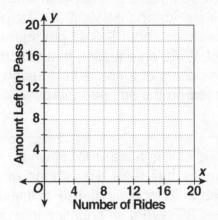

_____

_____

2. The rent charged for space in an office building is related to the size of the space rented. The rent for 600 square feet of floor space is $750, while the rent for 900 square feet is $1150. Write an equation for the rent y based on the square footage of the floor space x.

_____

Choose the letter of the correct answer.

3. A limousine charges $35 plus $2 per mile. Which equation shows the total cost of a ride in the limousine?

   A $y = 35x + 2$ **C** $y = 2x - 35$

   B $y = 2x + 35$ **D** $2x + 35y = 2$

4. A newspaper pays its delivery people $75 each day plus $0.10 per paper delivered. Which equation shows the daily earnings of a delivery person?

   F $y = 0.1x + 75$ **H** $x + 0.1y = 75$

   G $y = 75x + 0.1$ **J** $0.1x + y = 75$

5. A friend gave Ms. Morris a $50 gift card for a local car wash. If each car wash costs $6, which equation shows the number of dollars left on the card?

   A $50x + 6y = 1$ **C** $y = -6x + 50$

   B $y = 6x + 50$ **D** $y = 6x - 50$

6. Antonio's weekly allowance is given by the equation $A = 0.5c + 10$, where c is the number of chores he does. If he received $16 in allowance one week, how many chores did he do?

   F 10   **H** 14

   G 12   **J** 15

Holt Mathematics

LESSON 12-4

Problem Solving
Point-Slope Form

Write the correct answer.

1. A 1600 square foot home in Houston will sell for about $102,000. The price increases about $43.41 per square foot. Write an equation that describes the price *y* of a house in Houston, based on the square footage *x*.

2. Write the equation in Exercise 1 in slope-intercept form.

3. Wind chill is a measure of what temperature feels like with the wind. With a 25 mph wind, 40°F will feel like 29°F. Write an equation in point-slope form that describes the wind chill *y* based on the temperature *x*, if the slope of the line is 1.337.

4. With a 25 mph wind, what does a temperature of 0°F feel like?

From 2 to 13 years, the growth rate for children is generally linear. Choose the letter of the correct answer.

5. The average height of a 2-year old boy is 36 inches, and the average growth rate per year is 2.2 inches. Write an equation in point-slope form that describes the height of a boy *y* based on his age *x*.

   A $y - 36 = 2(x - 2.2)$

   B $y - 2 = 2.2(x - 36)$

   C $y - 36 = 2.2(x - 2)$

   D $y - 2.2 = 2(x - 36)$

6. The average height of a 5-year old girl is 44 inches, and the average growth rate per year is 2.4 inches. Write an equation in point-slope form that describes the height of a girl *y* based on her age *x*.

   F $y - 2.4 = 44(x - 5)$

   G $y - 44 = 2.4(x - 5)$

   H $y - 44 = 5(x - 2.4)$

   J $y - 5 = 2.4(x - 44)$

7. Write the equation from Exercise 6 in slope-intercept form.

   A $y = 2.4x - 100.6$

   B $y = 44x - 217.6$

   C $y = 5x + 32$

   D $y = 2.4x + 32$

8. Use the equation in Exercise 6 to find the average height of a 13-year old girl.

   F 56.3 in.

   G 63.2 in.

   H 69.4 in.

   J 97 in.

Holt Mathematics

Problem Solving

LESSON 12-5 *Direct Variation*

Determine whether the data sets show direct variation. If so, find the equation of direct variation.

1. The table shows the distance in feet traveled by a falling object in certain times.

Time (s)	0	0.5	1	1.5	2	2.5	3
Distance (ft)		0	4	16	36	64	100	144

_____

_____

2. The R-value of insulation gives the material's resistance to heat flow. The table shows the R-value for different thicknesses of fiberglass insulation.

Thickness (in)	1	2	3	4	5	6
R-value	3.14	6.28	9.42	12.56	15.7	18.84

_____

4. The table shows the relationship between degrees Celsius and degrees Fahrenheit.

3. The table shows the lifting power of hot air.

Hot Air (ft³)	50	100	500	1000	2000	3000
Lift (lb)	1	2	10	20	40	60

° Celsius	−10	−5	0	5	10	20	30
° Fahrenheit	14	23	32	41	50	68	86

_____

_____

The relationship between your weight on Earth and your weight on other planets is direct variation. The table below shows how much a person who weights 100 lb on Earth would weigh on the moon and different planets.

5. Find the equation of direct variation for the weight on earth *e* and on the moon *m*.

   A $m = 0.166e$ **C** $m = 6.02e$

   B $m = 16.6e$ **D** $m = 1660e$

Solar System Objects	Weight (lb)
Moon	16.6
Jupiter	236.4
Pluto	6.7

6. How much would a 150 lb person weigh on Jupiter?

   F 63.5 lb   **H** 354.6 lb

   G 286.4 lb   **J** 483.7 lb

7. How much would a 150 lb person weigh on Pluto?

   A 5.8 lb   **C** 12.3 lb

   B 10.05 lb   **D** 2238.8 lb

Holt Mathematics

Problem Solving

LESSON 12-6 *Graphing Inequalities in Two Variables*

The senior class is raising money by selling popcorn and soft drinks. They make $0.25 profit on each soft drink sold, and $0.50 on each bag of popcorn. Their goal is to make at least $500.

1. Write an inequality showing the relationship between the sales of *x* soft drinks and *y* bags of popcorn and the profit goal.

   _____

2. Graph the inequality from exercise 1.

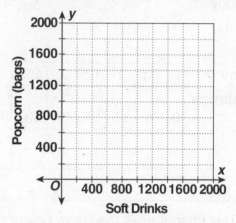

3. List three ordered pairs that represent a profit of exactly $500.

   _____

   _____

4. List three ordered pairs that represent a profit of more than $500.

   _____

   _____

5. List three ordered pairs that represent a profit of less than $500.

   _____

   _____

A vehicle is rated to get 19 mpg in the city and 25 mpg on the highway. The vehicle has a 15-gallon gas tank. The graph below shows the number of miles you can drive using no more than 15 gallons.

6. Write the inequality represented by the graph.

   A $\frac{x}{19} + \frac{y}{25} < 15$

   B $\frac{x}{19} + \frac{y}{25} \leq 15$

   C $\frac{x}{19} + \frac{y}{25} \geq 15$

   D $\frac{x}{19} + \frac{y}{25} > 15$

7. Which ordered pair represents city and highway miles that you can drive on one tank of gas?

   F (200, 150)   **H** (250, 75)

   G (50, 350)   **J** (100, 175)

8. Which ordered pair represents city and highway miles that you cannot drive on one tank of gas?

   A (100, 200)   **C** (50, 275)

   B (150, 200)   **D** (250, 25)

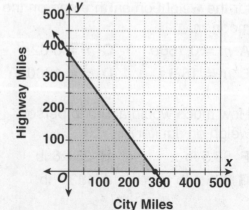

Holt Mathematics

Name _____ Date _____ Class _____

Problem Solving
Lines of Best Fit

Write the correct answer. Round to the nearest hundredth.

1. The table shows in what year different average speed barriers were broken at the Indianapolis 500. If x is the year, with $x = 0$ representing 1900, and y is the average speed, find the mean of the x- and y- coordinates.

_____

Barrier (mi/h)	Year	Average Speed (mi/h)
80	1914	82.5
100	1925	101.1
120	1949	121.3
140	1962	140.9
160	1972	163.0
180	1990	186.0

2. Graph the data from exercise 1 and find the equation of the line of best fit.

_____

_____

3. Use your equation to predict the year the 210 mph barrier will be broken.

_____

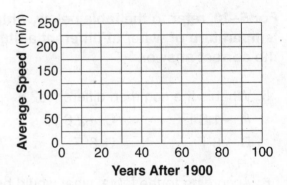

The percent of the U.S. population who smokes can be represented by the line of best fit with the equation $y = -0.57x + 44.51$ where x is the year, $x = 0$ represents 1960, and y is the percent of the population who smokes. Circle the letter of the correct answer.

4. Which term describes the percent of the population that smokes?

   A Increasing   C No change

   B Decreasing   D Cannot tell

5. Use the equation to predict the percent of smokers in 2005.

   F 13.16%   H 24.56%

   G 18.86%   J 41.66%

6. Use the equation to predict when the percent of smokers will be less than 15%.

   A 1996   C 2012

   B 2010   D 2023

7. Use the equation to predict the percent of smokers in 2010.

   F 10.31%   H 16.01%

   G 11.01%   J 12.71%

Holt Mathematics

Name _____ Date _____ Class _____

A section of seats in an auditorium has 18 seats in the first row. Each row has two more seats than the previous row. There are 25 rows in the section. Write the correct answer.

1. List the number of seats in the second, third and fourth rows of the section.

2. How many seats are in the 10th row?

3. How many seats are in the 15th row?

4. In which row are there 32 seats?

For 5–10, refer to the table below, which shows the boiling temperature of water at different altitudes. Choose the letter of the correct answer.

5. What is the common difference?
   A −1.8°F   **C** −2.8°F
   B 1.8°F   **D** 6°F

Altitude (thousands of feet)	Boiling point of water (°F)
1	210.2
2	208.4
3	206.6
4	204.8
5	203

6. According to the table, what would be the boiling point of water at an altitude of 10,000 feet?
   F 192.2°F   **H** 226.4°F
   G 194°F   **J** 228.2°F

7. According to the table, what would be the boiling point of water at an altitude of 15,000 feet?
   A 181.4°F   **C** 185°F
   B 183.2°F   **D** 235.4°F

8. Estimate the boiling point of water in Jacksonville, Florida, which has an elevation of 0 feet.
   F 0°F   **H** 212°F
   G 208.4°F   **J** 213.8°F

9. The highest point in the United States is Mt. McKinley, Alaska, with an elevation of 20,320 feet. Estimate the boiling point of water at the top of Mt. McKinley.
   A 172.4°F   **C** 244.4°F
   B 176°F   **D** 246.2°F

10. At which elevation will the boiling point of water be less than 150°F?
    F 28,000 ft   **H** 32,000 ft
    G 30,000 ft   **J** 35,000 ft

Holt Mathematics

Problem Solving

LESSON 13-2 *Terms of Geometric Sequences*

For Exercises 1–2, determine if the sequence could be geometric. If so, find the common ratio. Write the correct answer.

1. A computer that was worth $1000 when purchased was worth $800 after six months, $640 after a year, $512 after 18 months, and $409.60 after two years.

2. A student works for a starting wage of $6.00 per hour. She is told that she can expect a $0.25 raise every six months.

3. A piece of paper that is 0.01 inches thick is folded in half repeatedly. If the paper were folded 6 times, how thick would the result be?

4. A vacuum pump removes one-half of the air in a container with each stroke. How much of the original air is left in the container after 8 strokes?

For exercises 5–8, assume that the cost of a college education increases an average of 5% per year. Choose the letter of the correct answer.

5. If the in-state tuition at the University of Florida is $2256 per year, what will the tuition be in 10 years?
   A $3174.24
   B $3333.14
   C $3499.80
   D $3674.79

6. If it costs $3046 per year for tuition for a Virginia resident at the University of Virginia now, how much will tuition be in 8 years?
   F $4183.26
   G $4286.03
   H $4500.33
   J $4725.35

7. If it costs $25,839 per year in tuition to attend Northwestern University now, how much will tuition be in 5 years?
   A $31,407.47
   B $32,977.84
   C $37,965.97
   D $42,483.72

8. If you start attending Northwestern University in 5 years and attend for 4 years, how much will you spend in total for tuition?
   F $142,138.61
   G $135,370.12
   H $131,911.36
   J $169,934.88

Holt Mathematics

LESSON 13-3 Problem Solving
Other Sequences

A toy rocket is launched and the height of the rocket during its first four seconds is recorded. Write the correct answer.

1. Find the first differences for the rocket's heights.

   _____

2. Find the second differences.

   _____

3. Use the first and second differences to predict the height of the rocket at 5, 6, and 7 seconds.

Time (sec)	Height (ft)
0	0
1	176
2	320
3	432
4	512
5	
6	
7	

4. What is the maximum height of the rocket?

5. When will the rocket hit the ground?

For exercises 6–9, refer to the table below, which shows the number of diagonals for different polygons. Choose the letter for the correct answer.

6. What are the first differences for the diagonals?

   A 1, 1, 1, 1   **C** 2, 3, 4, 5

   B 3, 2, 0, 3, 7   **D** 2, 7, 14, 23

Polygon	Sides	Diagonals
Triangle	3	0
Quadrilateral	4	2
Pentagon	5	5
Hexagon	6	9
Heptagon	7	14

7. What are the second differences?

   F 1, 1, 1   **H** 5, 7, 8

   G 1, 2, 3, 4   **J** 7, 9, 11, 13

8. How many diagonals does a nonagon (9 sides) have?

   A 21

   B 24

   C 27

   D 32

9. Which rule will give the number of diagonals d for s sides?

   F $d = \dfrac{s(s+1)}{2}$

   G $d = (s-3)(s-2) - 1$

   H $d = \dfrac{s(s-3)}{2}$

   J $d = (s-3)(s-2)$

Holt Mathematics

Name _____ Date _____ Class _____

Write the correct answer.

1. The greatest amount of snow that has ever fallen in a 24-hour period in North America was on April 14–15, 1921 in Silver Lake, Colorado. In 24 hours, 76 inches of snow fell, at an average rate of 3.2 inches per hour. Find a rule for the linear function that describes the amount of snow after x hours at the average rate.

_____

2. At the average rate of snowfall from Exercise 1, how much snow had fallen in 15 hours?

_____

3. The altitude of clouds in feet can be found by multiplying the difference between the temperature and the dew point by 228. If the temperature is 75°, find a rule for the linear function that describes the height of the clouds with dew point x.

_____

4. If the temperature is 75° and the dew point is 40°, what is the height of the clouds?

_____

For exercises 5–7, refer to the table below, which shows the relationship between the number of times a cricket chirps in a minute and temperature.

5. Find a rule for the linear function that describes the temperature based on x, the number of cricket chirps in a minute based on temperature.

  A $f(x) = x + 5$

  B $f(x) = \frac{x}{4} + 40$

  C $f(x) = x - 20$

  D $f(x) = \frac{x}{2} + 20$

Cricket Chirps/min	Temperature (°F)
80	60
100	65
120	70
140	75

6. What is the temperature if a cricket chirps 150 times in a minute?

  F 77.5°F   **H** 130°F

  G 95°F   **J** 155°F

7. If the temperature is 85°F, how many times will a cricket chirp in a minute?

  A 61   **C** 180

  B 105   **D** 200

Holt Mathematics

Name _____ Date _____ Class _____

Problem Solving
Exponential Functions

From 1950 to 2000, the world's population grew exponentially. The function that models the growth is $f(x) = 1.056 \cdot 1.018^x$ where x is the year ($x = 50$ represents 1950) and $f(x)$ is the population in billions. Round each number to the nearest hundredth.

1. Estimate the world's population in 1950.

   _____

2. Estimate the world's population in 2005.

   _____

3. Predict the world's population in 2025.

   _____

4. Predict the world's population in 2050.

   _____

Insulin is used to treat people with diabetes. The table below shows the percent of an insulin dose left in the body at different times after injection.

5. Which ordered pair does not represent a half-life of insulin?

   A (24, 70.71)   **C** (48, 50)

   B (50, 50)   **D** (72, 35.35)

Time elapsed (min)	Percent remaining
0	100
48	50
96	25
144	12.5

6. Write an exponential function that describes the percent of insulin in the body after x half-lives.

   F $f(x) = 100\left(\frac{1}{2}\right)^x$ **H** $f(x) = 2(100)^x$

   G $f(x) = 10\left(\frac{1}{2}\right)^x$ **J** $f(x) = 48\left(\frac{1}{2}\right)^x$

7. What percent of insulin would be left in the body after 6 hours?

   A 0.25%   **C** 0.55%

   B 0.39%   **D** 1.56%

8. What percent of insulin would be left in the body after 9 hours?

   F 0.04%   **H** 0.17%

   G 0.12%   **J** 0.26%

9. A new form of insulin that is being developed has a half-life of 9 hours. Write an exponential function that describes the percent of insulin in the body after x half-lives.

   A $f(x) = 100\left(\frac{1}{2}\right)^x$ **C** $f(x) = 2(100)^x$

   B $f(x) = 9\left(\frac{1}{2}\right)^x$ **D** $f(x) = 100(9)^x$

10. What percent of the new form of insulin would be left in the body after 9 hours?

    F 12.5%   **H** 50%

    G 25%   **J** 75%

Holt Mathematics

LESSON 13-6

Problem Solving

Quadratic Functions

To find the time it takes an object to fall, you can use the equation $h = -16t^2 - vt + s$ where h is the height in feet, t is the time in seconds, v is the initial velocity, and s is the starting height in feet. Write the correct answer.

1. If a construction worker drops a tool from 240 feet above the ground, how many feet above the ground will it be in 2 seconds? Hint: $v = 0$, $s = 240$.

2. How long will it take the tool in Exercise 1 to hit the ground? Round to the nearest hundredth.

3. The Gateway Arch in St. Louis, Missouri is the tallest manmade memorial. The arch rises to a height of 630 feet. If you throw a rock down from the top of the arch with a velocity of 20 ft/s, how many feet above the ground will the rock be in 2 seconds?

4. Will the rock in exercise 3 hit the ground within 6 seconds of throwing it?

The average monthly rainfall for Seattle, Washington can be approximated by the equation $f(x) = 0.147x^2 - 1.890x + 7.139$ where x is the month (January: $x = 1$, February, $x = 2$, etc.) and $f(x)$ is the monthly rainfall in inches. Choose the letter for the best answer.

5. What is the average monthly rainfall in Seattle for the month of January?

    A 3.7 in   **C** 7.6 in

    B 5.4 in   **D** 9.2 in

6. What is the average monthly rainfall in Seattle for the month of April?

    F 0.2 in   **H** 1.9 in

    G 1.4 in   **J** 2.8 in

7. What is the average monthly rainfall in Seattle for the month of August?

    A 1.1 in   **C** 5.6 in

    B 1.4 in   **D** 6.8 in

8. In what month does it rain the least in Seattle, Washington?

    F May   **H** July

    G June   **J** August

Holt Mathematics

Problem Solving

13-7 Inverse Variation

For a given focal length of a camera, the f-stop varies inversely with the diameter of the lens. The table below gives the f-stop and diameter data for a focal length of 400 mm. Round to the nearest hundredth.

f-stop	diameter (mm)
1	400
2	200
4	100
8	50
16	25
32	12.5

1. Use the table to write an inverse variation function.

   _____

2. What is the diameter of a lens with an f-stop of 1.4?

   _____

3. What is the diameter of a lens with an f-stop of 11?

   _____

4. What is the diameter of a lens with an f-stop of 22?

The inverse square law of radiation says that the intensity of illumination varies inversely with the square of the distance to the light source.

5. Using the inverse square law of radiation, if you halve the distance between yourself and a fire, by how much will you increase the heat you feel?

   A 2

   B 4

   C 8

   D 16

6. Using the inverse square law of radiation, if you double the distance between a radio and the transmitter, how will it affect the signal intensity?

   F $\frac{1}{4}$ as strong

   G $\frac{1}{2}$ as strong

   H twice as strong

   J 4 times stronger

7. Using the inverse square law of radiation, if you increase the distance between yourself and a light by 4 times, how will it affect the light's intensity?

   A $\frac{1}{16}$ as strong   C $\frac{1}{2}$ as strong

   B $\frac{1}{4}$ as strong   D twice as strong

8. Using the inverse square law of radiation if you move 3 times closer to a fire, how much more intense will the fire feel?

   F $\frac{1}{3}$ as strong

   G 3 times stronger

   H 9 times stronger

   J 27 times stronger

Holt Mathematics

Problem Solving

LESSON 14-1 Polynomials

The table below shows expressions used to calculate the
surface area and volume of various solid figures where s is
side length, l is length, w is width, h is height, and r is radius.

1. List the expressions that are
   trinomials.

   _____

2. What is the degree of the expression
   for the surface area of a sphere?

   _____

3. A cube has side length
   of 5 inches. What is its surface area?

   _____

Solid Figure Polynomials

Solid Figure	Surface Area	Volume
Cube	$6s^2$	s^3
Rectangular Prism	$2lw + 2lh + 2wh$	lwh
Right Cone	$\pi rl + \pi r^2$	$\pi r^2 h$
Sphere	$4\pi r^2$	$\frac{4}{3}\pi r^3$

4. If you know the radius and height of
   a cone, you can use the expression
   $(r^2 + h^2)^{0.5}$ to find its slant height.
   Is this expression a polynomial?
   Why or why not?

   _____

   _____

5. If a sphere has a radius of 4 feet,
   what is its surface area and volume?
   Use $\frac{22}{7}$ for pi.

Circle the letter of the correct answer.

6. Which statement is true of all the
   polynomials in the volume column of
   the table?

   A They are trinomials

   B They are binomials.

   C They are monomials.

   D None of them are polynomials.

7. The height, in feet, of a baseball
   thrown straight up into the air from
   6 feet above the ground at 100 feet
   per second after t seconds is given
   by the polynomial $-16t^2 + 100t + 6$.
   What is the height of the baseball 4
   seconds after it was thrown?

   F 150 feet

   G 278 feet

   H 342 feet

   J 662 feet

Holt Mathematics

Problem Solving
LESSON 14-2 *Simplifying Polynomials*

Write the correct answer.

1. The area of a trapezoid can be found using the expression $\frac{h}{2}(b_1 + b_2)$ where h is height, b is the length of base$_1$, and b_2 is the length of base$_2$. Use the Distributive Property to write an equivalent expression.

2. The sum of the measures of the interior angles of a polygon with n sides is $180(n - 2)$ degrees. Use the Distributive Property to write an equivalent expression, and use the expression to find the sum of the measures of the interior angles of an octagon.

3. The volume of a box of height h is $2h^4 + h^3 + h^2 + h^2 + h$ cubic inches. Simplify the polynomial and then find the volume if the height of the box is 3 inches.

4. The height, in feet, of a rocket launched upward from the ground with an initial velocity of 64 feet per second after t seconds is given by $16(4t - t^2)$. Write an equivalent expression for the rocket's height after t seconds. What is the height of the rocket after 4 seconds?

Circle the letter of the correct answer.

5. The surface area of a square pyramid with base b and slant height l is given by the expression $b(b + 2l)$. What is the surface area of a square pyramid with base 3 inches and slant height 5 inches?

   A 13 square inches

   B 19 square inches

   C 39 square inches

   D 55 square inches

6. The volume of a box with a width of $3x$, a height of $4x - 2$, and a length of $3x + 5$ can be found using the expression $3x(12x^2 + 14x - 10)$. Which is this expression, simplified by using the Distributive Property?

   F $36x^2 + 42x - 30$

   G $15x^3 + 17x^2 - 7x$

   H $36x^3 + 14x - 10$

   J $36x^3 + 42x^2 - 30x$

Holt Mathematics

Name _____ Date _____ Class _____

Write the correct answer.

1. What is the perimeter of the quadrilateral?

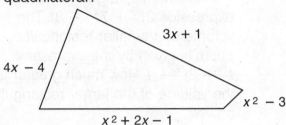

$3x + 1$

$4x - 4$

$x^2 - 3$

$x^2 + 2x - 1$

2. Jasmine purchased two rugs. One rug covers an area of $x^2 + 8x + 15$ and the other rug covers an area of $x^2 + 3x$. Write and simplify an expression for the combined area of the two rugs.

3. Anita's school photo is 12 inches long and 8 inches wide. She will surround the photo with a mat of width w. She will surround the mat with a frame that is twice the width of the mat. Find an expression for the perimeter of the framed photo.

4. The volume of a right cylinder is given by $\pi r^2 h$. The volume of a right cone is given by $\frac{1}{3}\pi r^2 h$. Write and simplify an expression for the total volume of a right cylinder and right cone combined, if the cylinder and cone have the same radius and height. Use 3.14 for π.

Choose the letter of the correct answer.

5. Each side of a square has length $4s - 2$. Which is an expression for the perimeter of the square.

   A $8s - 2$

   B $16s - 8$

   C $8s - 4$

   D $16s - 4$

6. The side lengths of a certain triangle can be expressed using the following binomials: $x + 3$, $2x + 2$, and $3x - 2$. Which is an expression for the perimeter of the triangle?

   F $2x + 5$

   G $2x - 1$

   H $3x + 5$

   J $6x + 3$

7. What polynomial can be added to $2x^2 + 3x + 1$ to get $2x^2 + 8x$?

   A $5x$

   B $5x + 1$

   C $5x^2 - 1$

   D $5x - 1$

8. Which of the following sums is NOT a binomial when simplified?

   F $(b^2 + 5b + 1) + (b^2 + 5b + 1)$

   G $(b^2 + 5b + 1) + (b^2 + 5b - 1)$

   H $(b^2 + 5b + 1) + (b^2 - 5b + 1)$

   J $(b^2 + 5b + 1) + (-b^2 + 5b + 1)$

Holt Mathematics

Write the correct answer.

1. Molly made a frame for a painting. She cut a rectangle with an area of $x^2 + 3x$ square inches from a piece of wood that had an area of $2x^2 + 9x + 10$ square inches. Write an expression for the area of the remaining frame.

_____

2. The volume of a rectangular prism, in cubic inches, is given by the expression $2t^3 + 7t^2 + 3t$. The volume of a smaller rectangular prism is given by the expression $t^3 + 2t^2 + t$. How much greater is the volume of the larger rectangular prism?

_____

3. The area of a square piece of cardboard is $4y^2 - 16y + 16$ square feet. A piece of the cardboard with an area of $2y^2 + 2y - 12$ square feet is cut out. Write an expression to show the area of the cardboard that is left.

_____

4. A container is filled with $3a^3 + 10a^2 - 8a$ gallons of water. Then $2a^3 - 3a^2 - 3a + 2$ gallons of water are poured out. How much water is left in the container?

_____

Circle the letter of the correct answer.

5. The perimeter of a rectangle is $4x^2 + 2x - 2$ meters. Its length is $x^2 + x - 2$ meters. What is the width of the rectangle?

A $3x^2 + x + 2$ meters

B $2x^2 + 2$ meters

C $x^2 + 1$ meters

D $\frac{3}{2}x - \frac{1}{2}x + 1$ meters

6. On a map, points A, B, and C lie in a straight line. Point A is $x^2 + 2xy + 5y$ miles from Point B. Point C is $3x^2 - 5xy + 2y$ miles from Point A. How far is Point B from Point C?

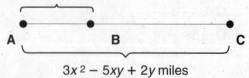

F $-2^2 + 7 + 3y$ miles

G $4x^2 - 3xy + 7y$ miles

H $-4x^2 + 3xy - 7y$ miles

J $2x^2 - 7xy - 3y$ miles

Holt Mathematics

Name _____ Date _____ Class _____

Write the correct answer.

1. A rectangle has a width of $5n^2$ inches and a length of $3n^2 + 2n + 1$ inches. Write and simplify an expression for the area of the rectangle. Then find the area of the rectangle if $n = 2$ inches.

_____

_____

2. The area of a parallelogram is found by multiplying the base and the height. Write and simplify an expression for the area of the parallelogram below.

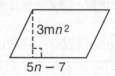

_____

_____

3. A parallelogram has a base of $2x^2$ inches and a height of $x^2 + 2x - 1$ inches. Write an expression for the area of the parallelogram. What is the area of the parallelogram if $x = 2$ inches?

_____

_____

4. A rectangle has a length of $x^2 + 2x - 1$ meters and a width of x^2 meters. Write an expression for the area of the rectangle. What is the area of the rectangle if $x = 3$ meters?

_____

_____

Circle the letter of the correct answer.

5. A rectangle has a width of $3x$ feet. Its length is $2x + \frac{1}{6}$ feet. Which expression shows the area of the rectangle?

   A $5x + \frac{1}{6}$

   B $6x^2 + \frac{1}{2}x^2$

   C $6x^2 + \frac{1}{2}$

   D $6x^2 + \frac{1}{2}x$

6. Which expression shows the area of the shaded region of the drawing?

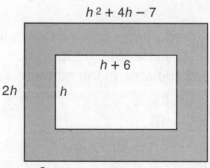

   F $2h^3 + 8h - 14h$

   G $2h^3 + 9h^2 - 8h$

   H $2h^3 + 7h^2 - 20h$

   J $2h^3 + 7h^2 - 8h$

Holt Pre-Algebra

LESSON 14-6 Problem Solving
Multiplying Binomials

Write and simplify an expression for the area of each polygon.

	Polygon	Dimensions	Area
1.	rectangle	length: $(n + 5)$; width: $(n - 4)$	
2.	rectangle	length: $(3y + 3)$; width: $(2y - 1)$	
3.	triangle	base: $(2b - 5)$; height: $(b^2 + 2)$	
4.	square	side length: $(m + 13)$	
5.	square	side length: $(2g - 4)$	
6.	circle	radius: $(3c + 2)$	

Choose the letter of the correct answer.

7. A photo is 8 inches by 11 inches. A frame of width x inches is placed around the photo. Which expression shows the total area of the frame and photo?

  A $x^2 + 19x + 88$

  B $4x^2 + 38x + 88$

  C $8x + 38$

  D $4x + 19$

8. Three consecutive odd integers are represented by the expressions, x, $(x + 2)$ and $(x + 4)$. Which expression gives the product of the three odd integers?

  F $x^3 + 8$

  G $x^3 + 6x^2 + 8x$

  H $x^3 + 6x^2 + 8$

  J $x^3 + 2x^2 + 8x$

9. A square garden has a side length of $(b - 4)$ yards. Which expression shows the area of the garden?

  A $2b - 8$

  B $b^2 + 16$

  C $b^2 - 8b - 16$

  D $b^2 - 8b + 16$

10. Which expression gives the product of $(3m + 4)$ and $(9m - 2)$?

  F $27m^2 + 30m - 8$

  G $27m^2 + 42m - 8$

  H $27m^2 + 42m + 8$

  J $27m^2 + 30m + 8$

Holt Mathematics